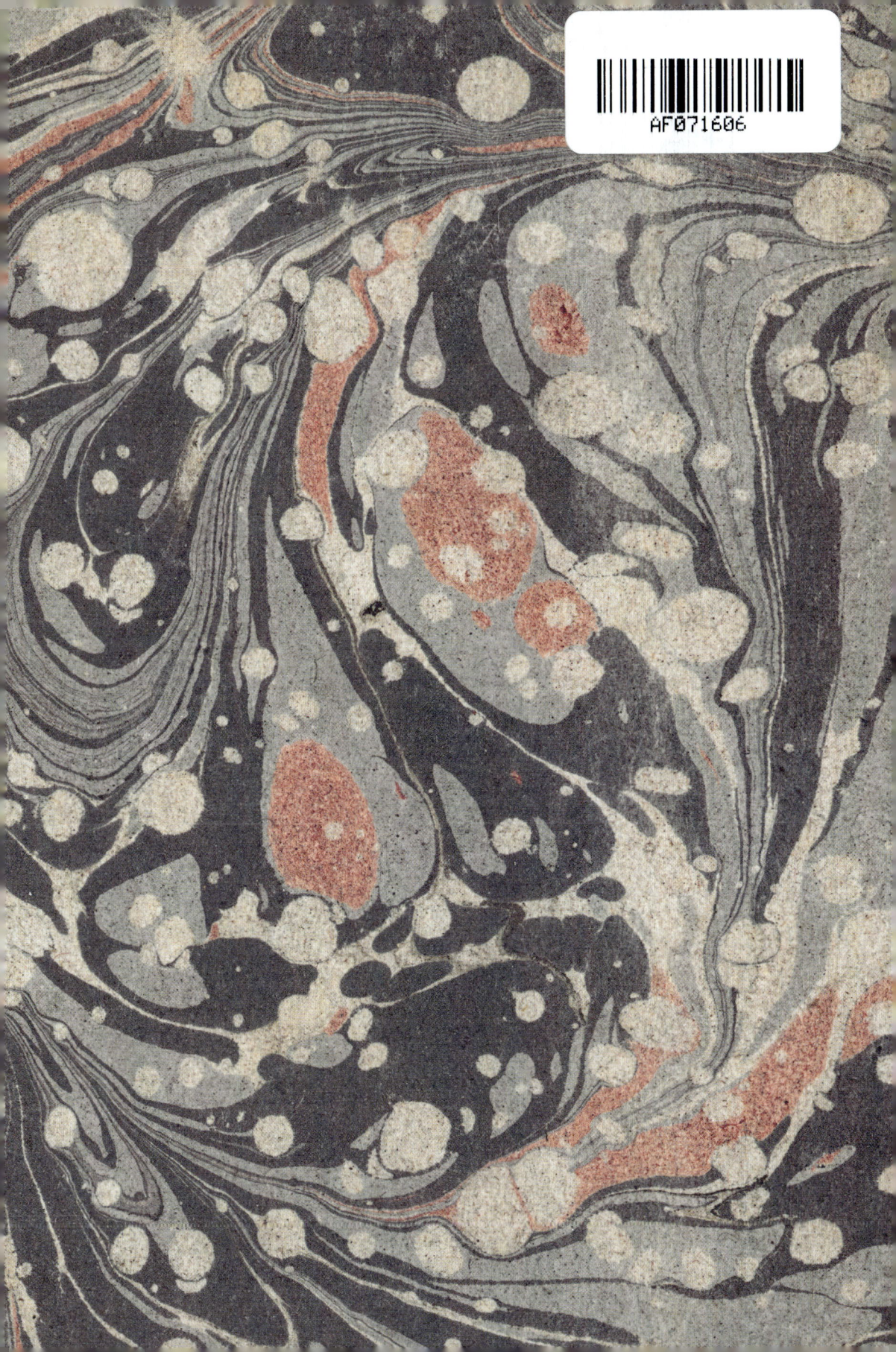

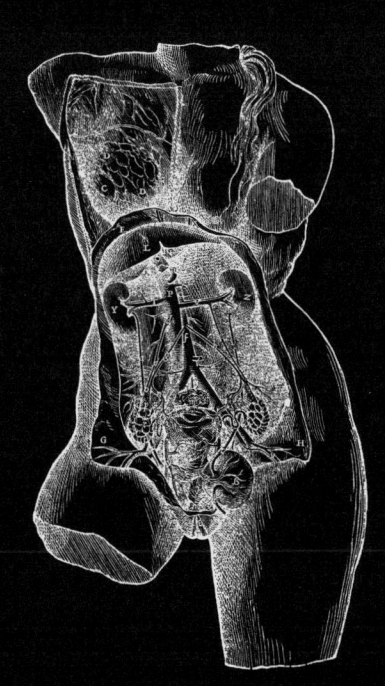

MORBID ANATOMY PRESENTS

THE ANATOMICAL VENUS

JOANNA·EBENSTEIN

with 366 illustrations, 324 in colour

FIRST PUBLISHED IN THE UNITED KINGDOM IN 2016 BY
THAMES & HUDSON LTD, 181A HIGH HOLBORN,
LONDON WC1V 7QX

REPRINTED 2022

The Anatomical Venus © 2016 THAMES & HUDSON LTD, LONDON

TEXT © 2016 JOANNA EBENSTEIN

DESIGN BY DANIEL STREAT AT VISUAL FIELDS

ALL RIGHTS RESERVED. NO PART OF THIS PUBLICATION
MAY BE REPRODUCED OR TRANSMITTED IN ANY FORM OR
BY ANY MEANS, ELECTRONIC OR MECHANICAL, INCLUDING
PHOTOCOPY, RECORDING OR ANY OTHER INFORMATION
STORAGE AND RETRIEVAL SYSTEM, WITHOUT PRIOR
PERMISSION IN WRITING FROM THE PUBLISHER.

BRITISH LIBRARY CATALOGUING-IN-PUBLICATION DATA
A CATALOGUE RECORD FOR THIS BOOK
IS AVAILABLE FROM THE BRITISH LIBRARY

ISBN 978-0-500-25218-5

PRINTED AND BOUND IN CHINA
BY EVERBEST PRINTING CO. LTD

BE THE FIRST TO KNOW ABOUT OUR
NEW RELEASES, EXCLUSIVE CONTENT
AND AUTHOR EVENTS BY VISITING
thamesandhudson.com
thamesandhudsonusa.com
thamesandhudson.com.au

[COVER] Venus Endormie *(Sleeping Beauty), Spitzner collection, courtesy of Université de Montpellier, collections anatomiques. Photo © Marc Dantan.*
[BACK COVER] *Anatomical Venus complete and deconstructed, Spitzner collection, courtesy of Université de Montpellier, collections anatomiques. Photo © Marc Dantan.*
[PAGE 1] *Female torso demonstrating the reproductive system, from the revolutionary anatomical atlas,* De Humani Corporis Fabrica *(On the Fabric of the Human Body), by Flemish anatomist Andreas Vesalius (1543).*
[PAGE 2] *Wax anatomical model of a woman in the ninth month of pregnancy, from the Berlin workshop of Gustav Zeiller, c. 1880.*
[PAGES 4–5] *Venerina (Little Venus), life-sized dissectible wax model created by the workshop of Clemente Susini at Florence's La Specola for Museo di Palazzo Poggi, Bologna, Italy (1782).*
[PAGES 6–7] *Wax anatomical models created by the workshop at La Specola, Florence, 1781–86, Josephinum Museum, Vienna, Austria.*
[PAGES 8–9] *Sleeping Beauty, a breathing waxwork modelled by Dr Philippe Curtius in Paris (1767).*
[PAGES 10–11] *Wax Anatomical Venus, intact and in partial dissection, by the workshop of Rudolf Pohl, Dresden, Germany (c. 1930).*
[PAGE 12] *Life-sized plaster anatomical model, blindfolded for storage, Blythe House, London (c. 1900).*

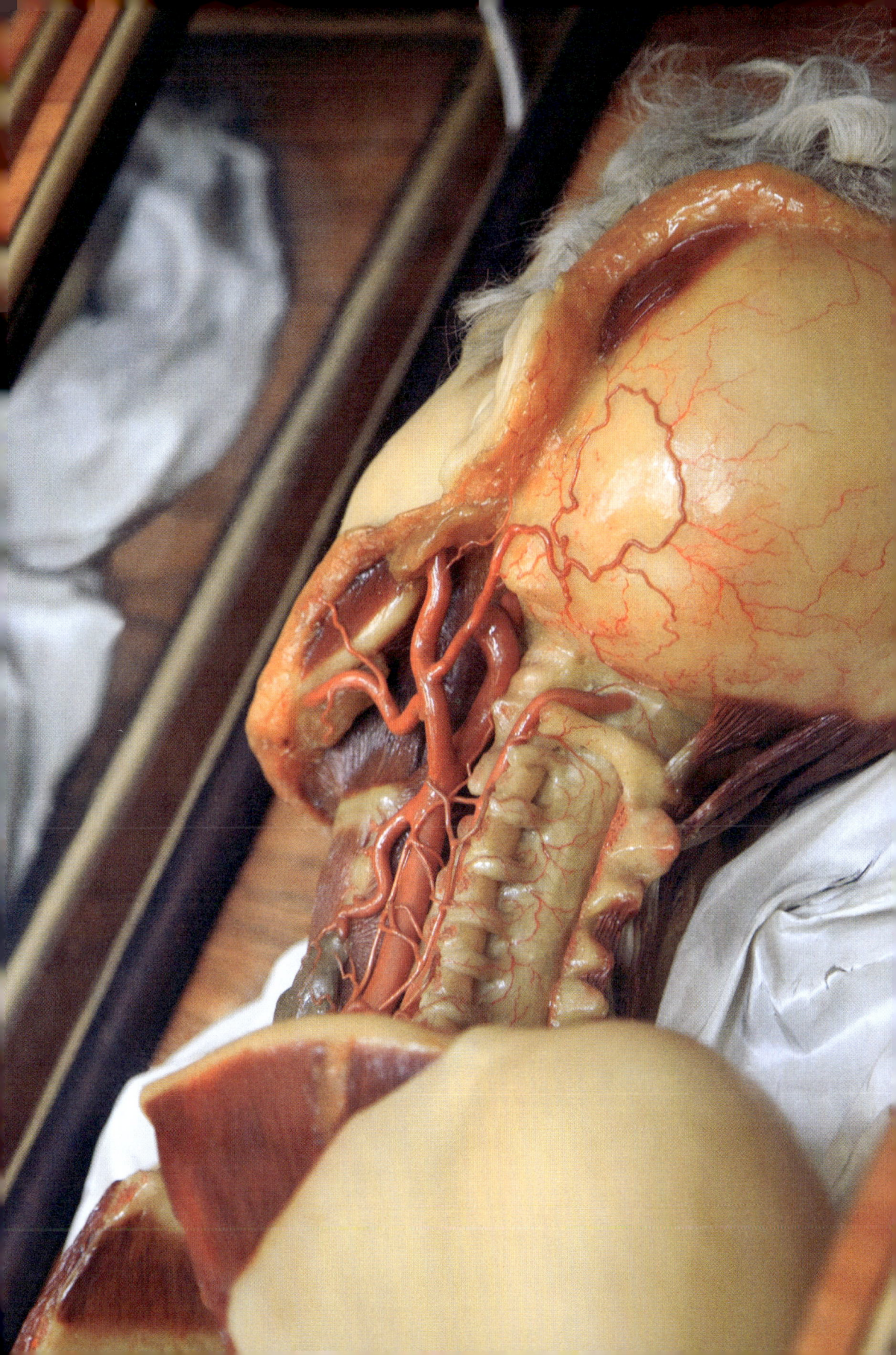

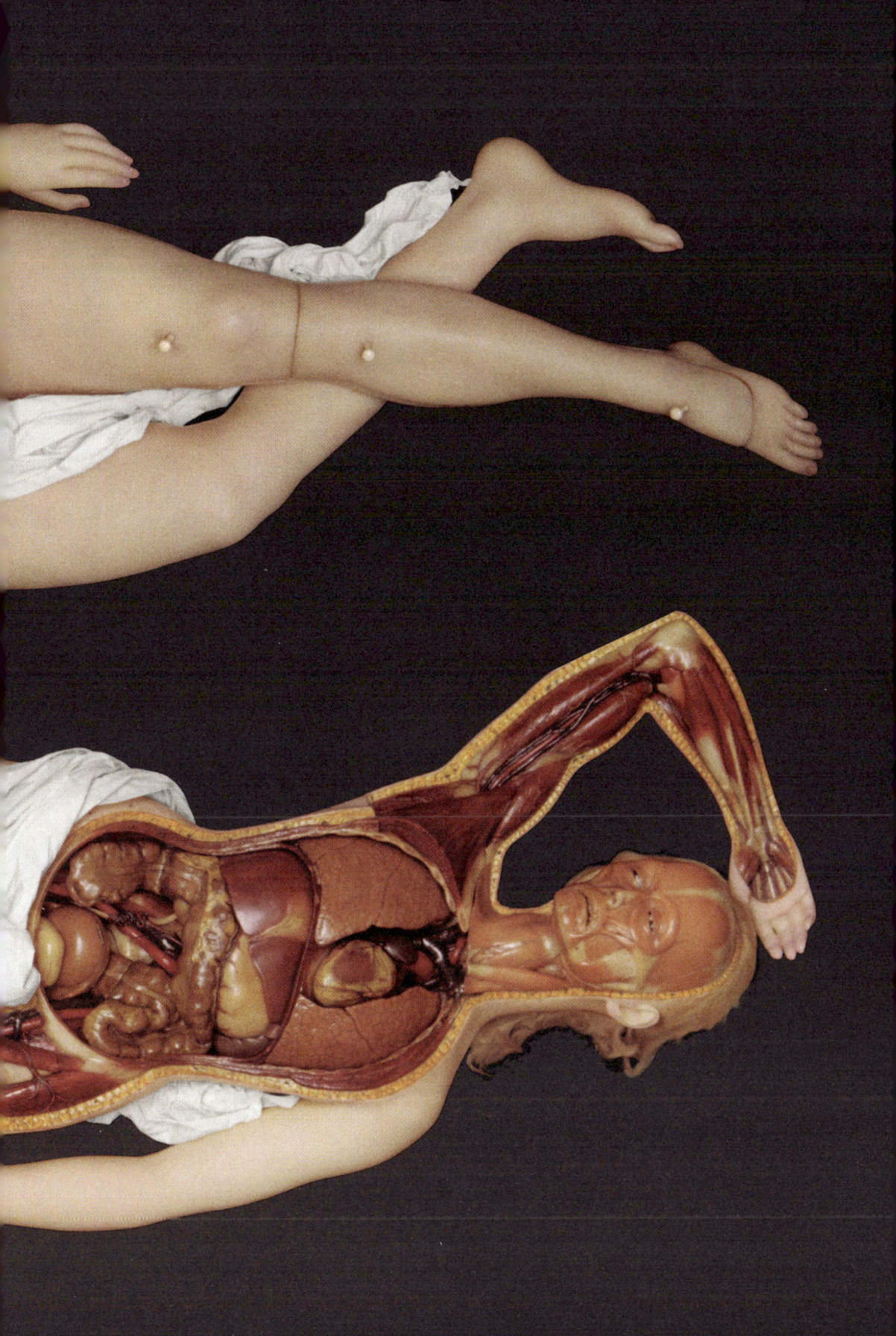

CONTENTS

INTRODUCTION:

UNVEILING·AN·ANATOMICAL·ENIGMA (14)

(1) THE·BIRTH·OF·THE·ANATOMICAL·VENUS (20)

(2) FROM·SACRED·TO·SCIENTIFIC·USE·OF·WAX (66)

(3) VENUS·AT·THE·FAIRGROUND (118)

(4) ECSTASY,·FETISHISM,·AND·DOLL·WORSHIP (178)

(5) VENUS,·THE·UNCANNY,·AND·THE·GHOST

IN·THE·MACHINE (200)

PLACES·OF·INTEREST (216)

SELECT·BIBLIOGRAPHY (218)

PICTURE·CREDITS (220)

INDEX (222)

ACKNOWLEDGMENTS (224)

UNVEILING·AN ANATOMICAL·ENIGMA

The purpose of anatomical images during the period of the Renaissance to the nineteenth century had as much to do with what we would call aesthetic and theological understanding as with the narrower interests of medical illustrators as now understood.... They were not simply instructional diagrams for the doctor technician, but statements about the nature of human beings as made by God in the context of the created world as a whole...they are about the nature of life and death... — MARTIN KEMP & MARINA WALLACE, **SPECTACULAR BODIES** (2000)

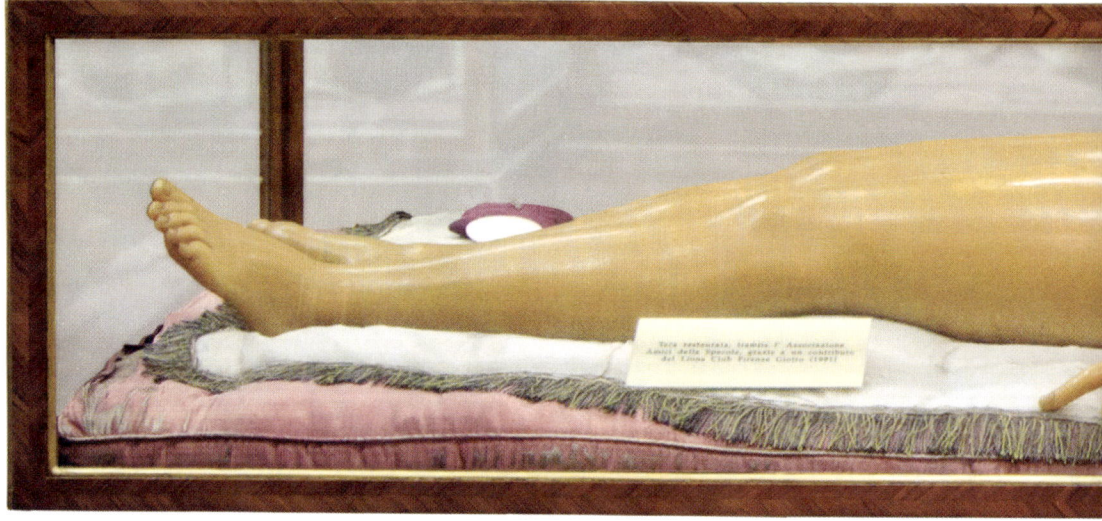

fig. 1

fig. 1 The most iconic dissectible wax Anatomical Venus— also known as the 'Demountable Venus' and the 'Medici Venus'— from the workshop of Clemente Susini at La Specola, Florence, Italy (1780–82). Life-sized.

Clemente Susini's Anatomical Venus, created 1780–82, is the perfect object: one whose luxuriously bizarre existence challenges belief. It—or better, she—was conceived as a means of teaching human anatomy without the need for constant dissection, which was messy, ethically fraught, and reliant on scarce cadavers. The Anatomical Venus also tacitly communicated the relationship between the human body and a divinely created cosmos, between art and science, and between nature and mankind as it was then understood.

Often referred to as the 'Medici Venus' or the 'Demountable Venus', this life-sized, dissectible wax woman with gleaming glass eyes and human hair can still be viewed in her original Venetian glass and rosewood case. She can be disassembled into seven anatomically correct layers, revealing at the final remove a tranquil fetus curled in her womb. She and her sisters, wax women in fixed states of anatomical undress sometimes referred to as Slashed Beauties or Dissected Graces, can still be found in a handful of European museums. Supine in their glass boxes, they beckon with a gentle smile or an ecstatic downcast gaze. One idly toys with a plait of real golden human hair; another clutches at the plush, moth-eaten satin cushions of her case as her torso erupts in a spontaneous, bloodless auto-dissection; another is crowned with a golden tiara, while one further wears a silk ribbon tied in a bow around a dangling entrail.

Since their creation in late-eighteenth-century Florence, these wax women have seduced, intrigued, and instructed. In the twenty-first century, they also confound, flickering on the edges of medicine and myth, votive and vernacular, fetish and fine art. How can we understand today an object that is at once a seductive representation of ideal female beauty and an explicit demonstration of the inner workings of the body? How can we make sense of an artefact that was once equally at home in the fairground and the medical museum? How

OVERLEAF
Illustrations of the anatomized female body from the fifteenth to nineteenth centuries. Examples include 1670 paper anatomies with moveable flaps to mimic real dissections (left page, middle row) and a figure half skeleton, half lady of fashion, standing next to an obelisk inscribed with biblical quotations (right page, bottom row, centre), from Life & Death Contrasted, or, an Essay on Woman *(c. 1770).*

can we comprehend a creature memorably described by Holly Myers in the *Los Angeles Times* as 'an Enlightenment-era St Teresa ravished by communion with the invisible forces of science'?

This book explores the contradictions inherent in the Anatomical Venus and attempts to answer these questions. Drawing on the scholarship of a broad array of medical and art historians, cultural theorists, and philosophers, this book contextualizes the Anatomical Venus, examining the beliefs and practices that led to her creation and revealing how she was received by her contemporary audience. It goes on to investigate the very different ways she came to be framed and perceived in the nineteenth century, and finally to trace her curious after-lives in the twentieth and twenty-first centuries.

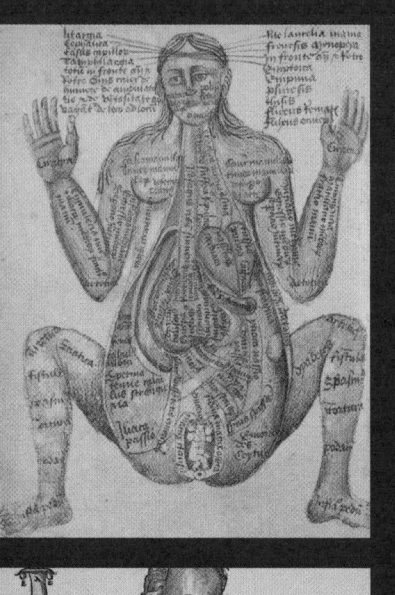
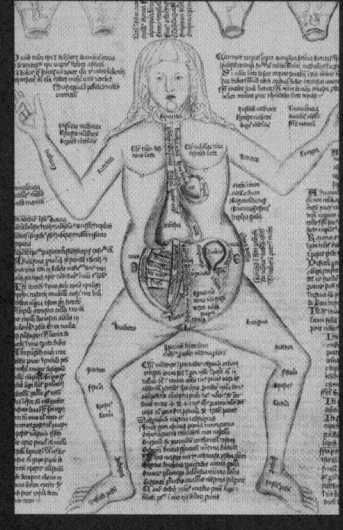
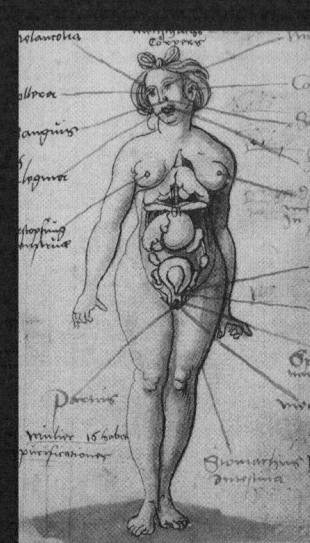
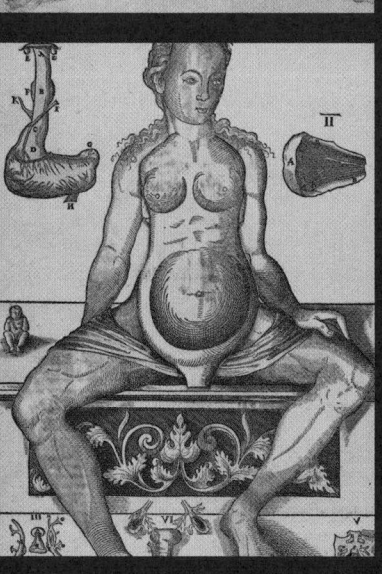
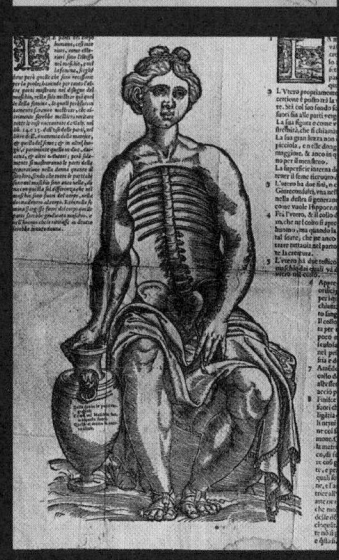

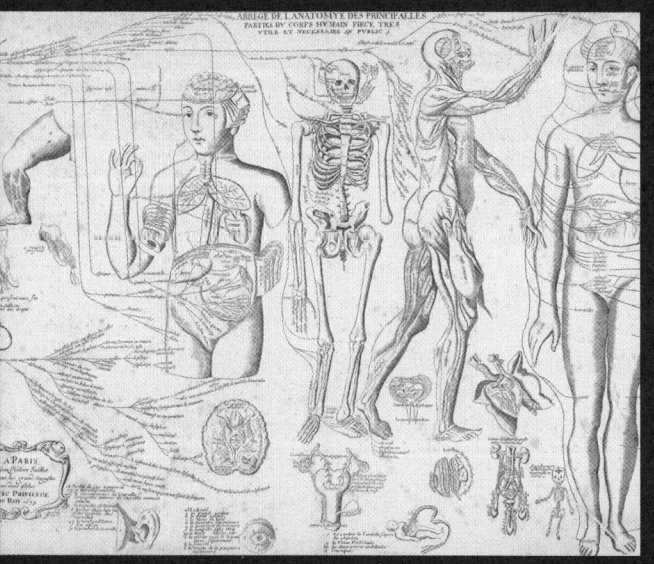
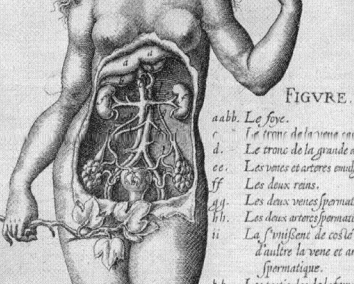

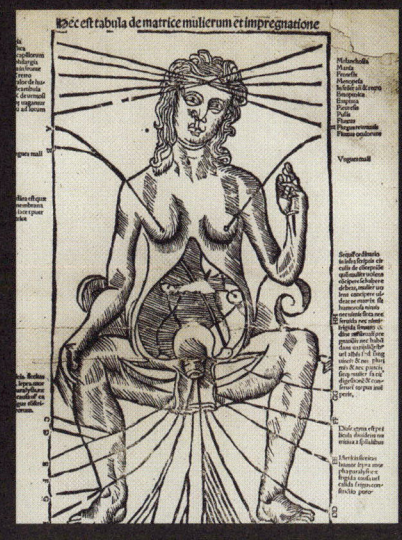

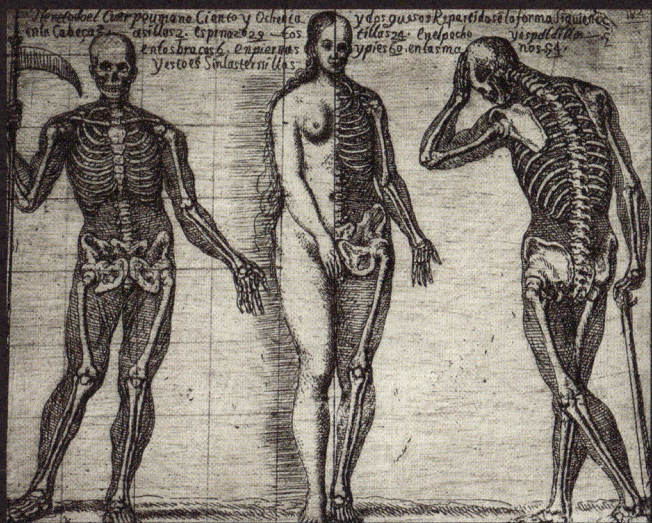
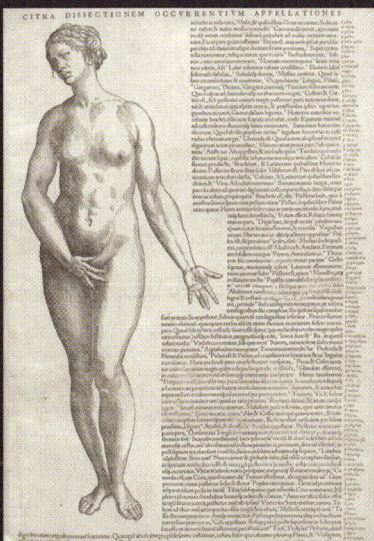

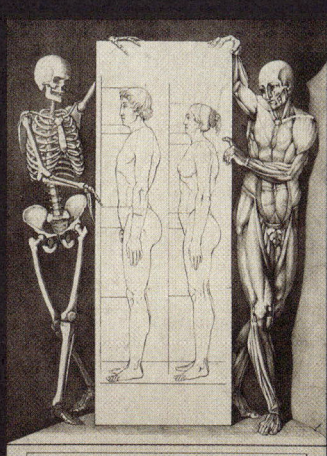

INTRODUCTION

The Anatomical Venus begins at the time and place of the creation of the Medici Venus—eighteenth-century Florence, Italy. Staunchly Catholic, with a long tradition of realistic wax anatomies in the form of sacred ex-votos manufactured for the pilgrims trade, Florence was also the epicentre of the Renaissance, which ushered in a great flowering of naturalistic representation in the arts and an unprecedented popular interest in the study of human anatomy. Ruled and patronized by the Medici dynasty from the fifteenth to the eighteenth century, in 1765 it came under the power of Leopold II (1747–92), a secular and humanitarian ruler from Habsburg, Vienna. Ten years later, led by his Enlightenment values, Leopold II—as Grand Duke of Tuscany—founded the first truly public science museum in Florence in 1775. Under the leadership of the court's physician, natural philosopher Felice Fontana (1730–1805), the wax workshop at the

fig. 2 Early-eighteenth-century dissectible ivory miniature manikin of a pregnant female, Italy.

fig. 2

fig. 3

fig. 3 Eighteenth-century wooden and ivory miniature based on Rembrandt's famous painting The Anatomy Lesson of Dr Nicolaes Tulp (1632). The original subject—Aris Krindt, a man hanged for robbery on 16 June 1632 and dissected at the Surgeons' Guild, Amsterdam—has been replaced by a dissectible female figure, similar to the ivory anatomical manikins of the period (see pp. 36, 52–53).

museum's core attempted to create an encyclopaedia of the human body in wax with which to teach and delight a popular audience. The finest and most iconic of these wax bodies—and the centrepiece of the museum—was the Medici Venus.

This book examines the precursors of the Medici Venus—anatomized and dissectible female figures in both sacred and profane domains—along with the artistic and theological predecessors that provided an aesthetic and conceptual framework for these models of feminine beauty intended to beguile as well as instruct. It tells the story of the creation of the first anatomical museum in Bologna, equal parts church and pedagogical tool, founded by the scientifically minded Pope Benedict XIV (1675–1758). It also investigates the similar yet conflicting ways in which Catholicism and medicine seek to preserve and effigize the body, and delves into the history of wax itself, uncannily similar in appearance to human flesh, and used since ancient times for a variety of death- and magic-related purposes. The book follows the journey of the Anatomical Venus to the fairground and the popular museum of the nineteenth century, where she and her sisters—living, wax and even automated 'sleeping beauties' in glass boxes—served as alluring centrepieces for displays that were at once educational and titillating. It traces the transformation in the meaning of the ecstatic from a religious, mystical experience to an erotic one, and follows the degeneration of the Anatomical Venus from beautiful instructional model to passive, life-sized doll created for men who prefer idealized surrogates to real women, or who have fashioned effigies of their beloveds in order to possess them for all time. It also looks at some of the ways in which artists and writers have taken the Anatomical Venus as their departure point or muse.

Finally, *The Anatomical Venus* provokes reflections upon the ways in which formerly mystical experiences have been sublimated and in which the ghost has officially become redundant in the machine. It ultimately considers why

fig. 4

fig. 5

fig. 4 Miniature wax memento-mori models depicting a fashionable Regency-era man and woman with semi-exposed skeletons. UK (c. 1800).

the Anatomical Venus has come to seem so strange to modern sensibilities, a classic example of the uncanny. Only a little over two hundred years ago she was the perfect tool to teach human anatomy to the public; today she is bizarre—an alluring, life-like female wax model in a state of ambiguous ecstasy with her inner organs on graphic display. Perhaps she could only be truly understood for a brief period, a time when it was still possible for religion, art, philosophy, and science to coexist peacefully; she is a relic from an age in which the torch was passed from spirituality to science as the primary arbiter of death, disease, the nature of life, and humanity's place in the universe. In her passive, waxen external beauty and realistically represented innards we can perceive a lost attitude to life: one that unifies rather than divides and allows for mystery and incomprehension. This book describes the enigma that makes the Anatomical Venus so fascinating without seeking to destroy it. It investigates her function, beauty, and evolving forms and uses without spoiling her charm, casting a wistful look back at a time when the study of nature was also the study of philosophy.

fig. 5 Two of nine wax plaques demonstrating female anatomy and fetal development. Probably made in Vienna, Austria (c. 1801–30).

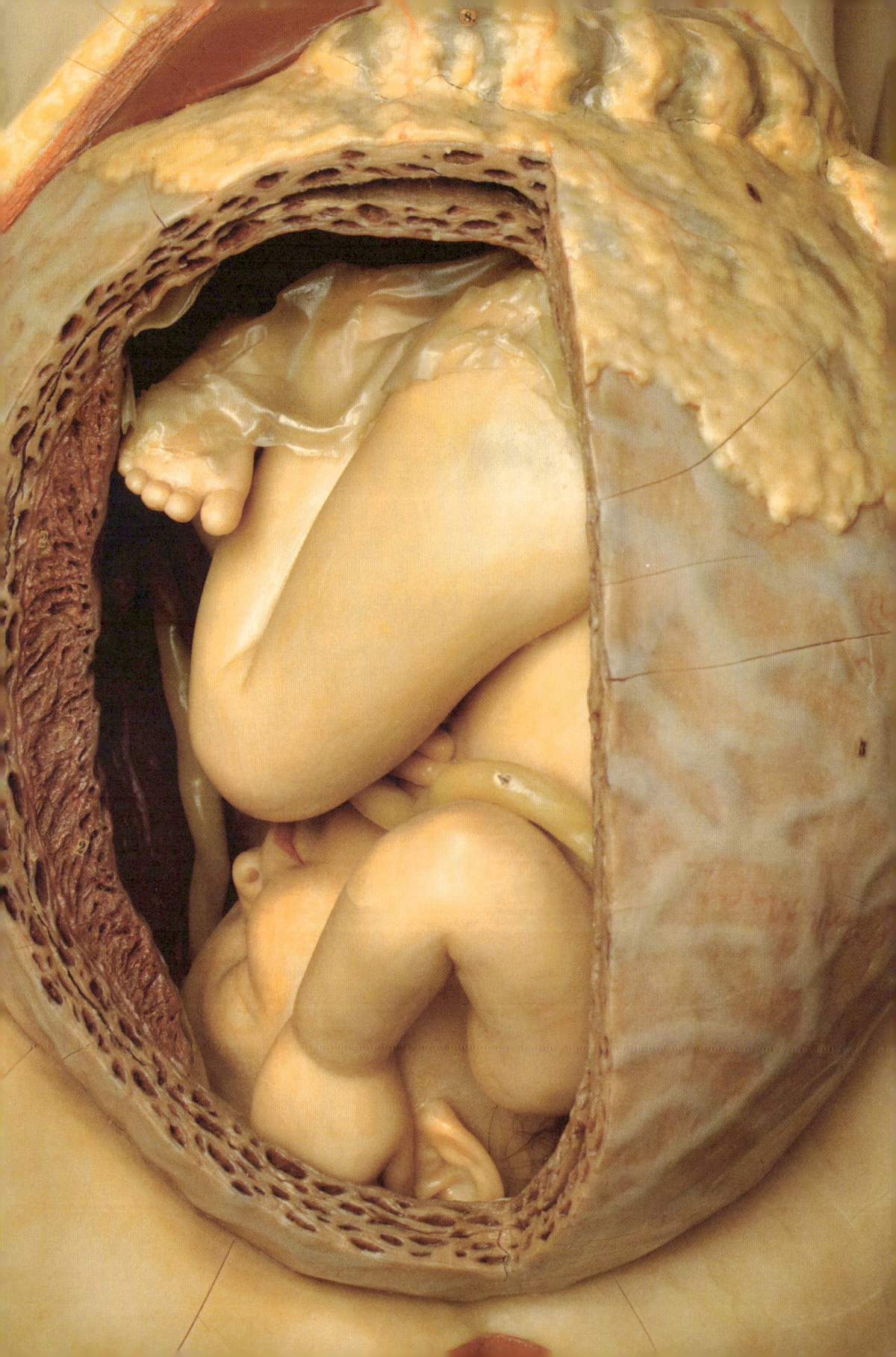

CHAPTER ONE

THE·BIRTH OF·THE ANATOMICAL VENUS

(21)

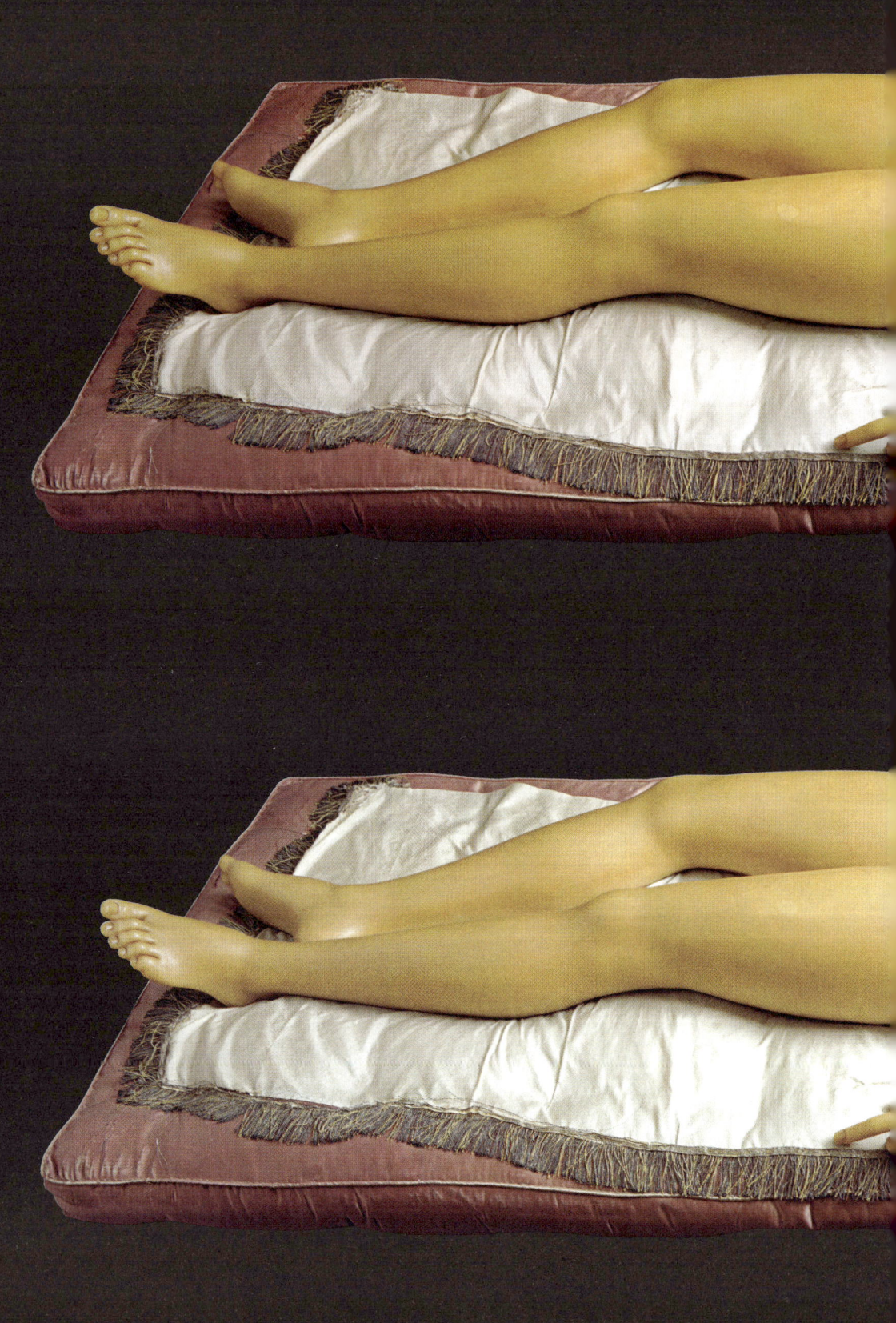

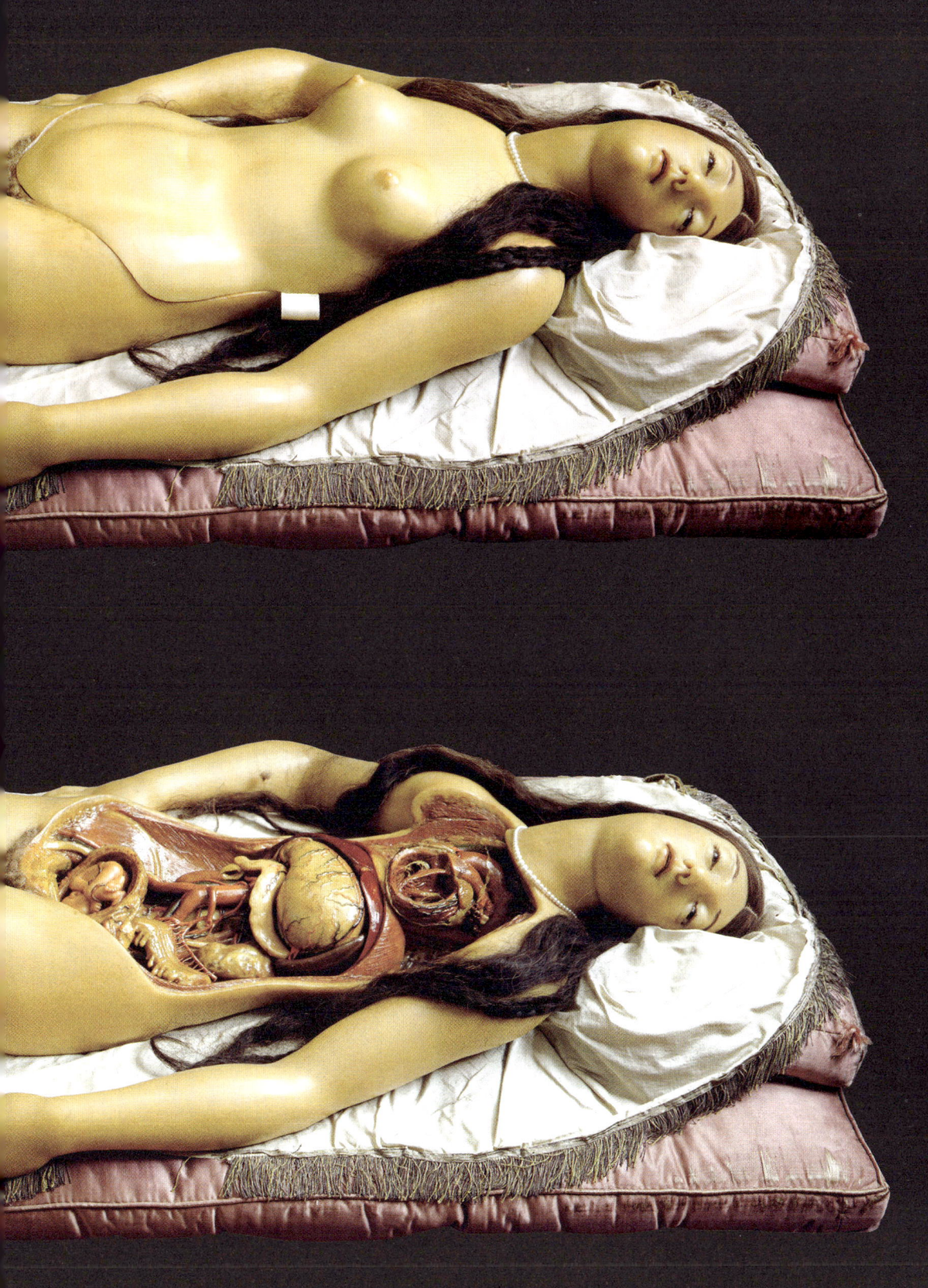

PREVIOUS
page 20
Detail of a wax model in the ninth month of pregnancy. From the workshop of Gustav Zeiller, Berlin, Germany (c. 1880).

pages 22–23
The Medici Venus (1780–82), displayed intact and dissected.

Towards the end of the eighteenth century, in a wax workshop in Florence, a life-sized, anatomically correct, dissectible goddess of coloured wax was created. Artist and master ceroplastician Clemente Susini (1754–1814) took the idealized feminine beauty for which Italian artists had long been renowned in an ambitious new direction, and to hyper-realistic lengths. The result—an Anatomical Venus known as the 'Medici Venus' or the 'Demountable Venus'—is a masterwork of human ingenuity; the product of a mystic marriage between art, science, and metaphysics.

The Medici Venus swoons langorously, apparently in the full flush of health, on a silk cushion in a casket of fine wood and Venetian glass. She is designed

fig. 6

fig. 7

fig. 6 Wax self-portrait of Clemente Susini (1754–1814), who oversaw the creation of the finest wax models at La Specola.

fig. 7 Marble bust of Felice Fontana (1730–1805), natural philosopher and director of the museum and wax workshop at La Specola. Fontana strictly monitored his employees and resented state supervision, refusing to keep proper records.

to charm in every detail: her glistening glass eyes are rimmed with real eyelashes, her bared throat is bound by a string of pearls, and she boasts a lustrous cascade of human hair. Seemingly alive but for her stillness and supernatural perfection, if you lift off her breastplate you will find that she is completely dissectible into seven layers, each revealing perfectly rendered, anatomically accurate organs. At the final remove—despite the figure betraying no outward signs of pregnancy—you will find a perfect, tranquil fetus curled in her womb: the *raison d'être* of the female body, at the the time.

To the modern eye, the Medici Venus is a perplexing object, one that challenges conventions of scientific visualization and explodes neat categorical divides between art and science, entertainment and education. In her own day, she was considered the ideal tool to teach anatomy to a general public, alleviating the need to resort to actual cadavers; indeed, she was so highly regarded by anatomists that copies were commissioned for a variety of museums and teaching collections around Europe. The Medici Venus was a perfect embodiment of the Enlightenment values of her time, in which human anatomy was understood as a reflection of the world and the pinnacle of divine knowledge, and in which to know the human body was to know the mind of God. Although she was neither the first nor the last of her kind, she was the most accomplished Anatomical Venus ever made, setting the standard by which all other Anatomical Venuses—or reclining anatomized wax women—are today judged.

The Medici Venus was born in the workshop of The Museum for Physics and Natural History in Florence, better known as La Specola (after a new observatory was added in 1789). The museum was founded by Leopold II, a revolutionary new leader from Vienna's Habsburg royal family. He became Grand Duke of Tuscany in 1765, succeeding his father, who had inherited the territory when the last Medici, Gian Gastone, died without an heir, ending three centuries of Medici dynastic rule. The Florence he inherited was far from

fig. 8

fig. 9

the centre of wealth and influence the city had been during the Renaissance. Determined to address Tuscany's decline and what he regarded as the more irrational practices of the Roman Catholic church, Leopold II set about a programme of social and economic reform based upon his own progressive principles. He abolished corporal punishment, executions, and torture; ended inquisitional courts and prisons; established health care for the poor; and forgave the public debt. In a radical departure from his predecessors, Leopold II believed himself to govern by social contract rather than by divine or sovereign right.

Leopold II's decision to open a public science museum in Florence was a central part of his Enlightenment mission to turn his new subjects into 'citizens' by educating them in the empirical observation of natural laws. His new museum would make available to the general public the rare and valuable cultural artefacts previously secreted in the Medici *Wunderkammern*, or cabinets of wonder. *Wunderkammern*, the precursors of today's museums, were private collections filled with the wonders of the world, both *naturalia* (natural objects) and *artificialia* (man-made objects). At this time, and up until the nineteenth century, science as we now understand it did not yet exist; the study of the natural world was largely the province of natural philosophy, which incorporated a variety of

fig. 8 Detail of an eighteenth-century portrait of the popular and learned 'Enlightenment Pope', Pope Benedict XIV (1675–1758).

fig. 9 Detail of a portrait of Leopold II (1747–92), Grand Duke of Tuscany, after he became Holy Roman Emperor in 1790.

Clemente Susini's dissectible Medici Venus (1780–82). Female anatomical figures almost always have their skin intact, while male figures are more likely to be depicted flayed to their muscles.

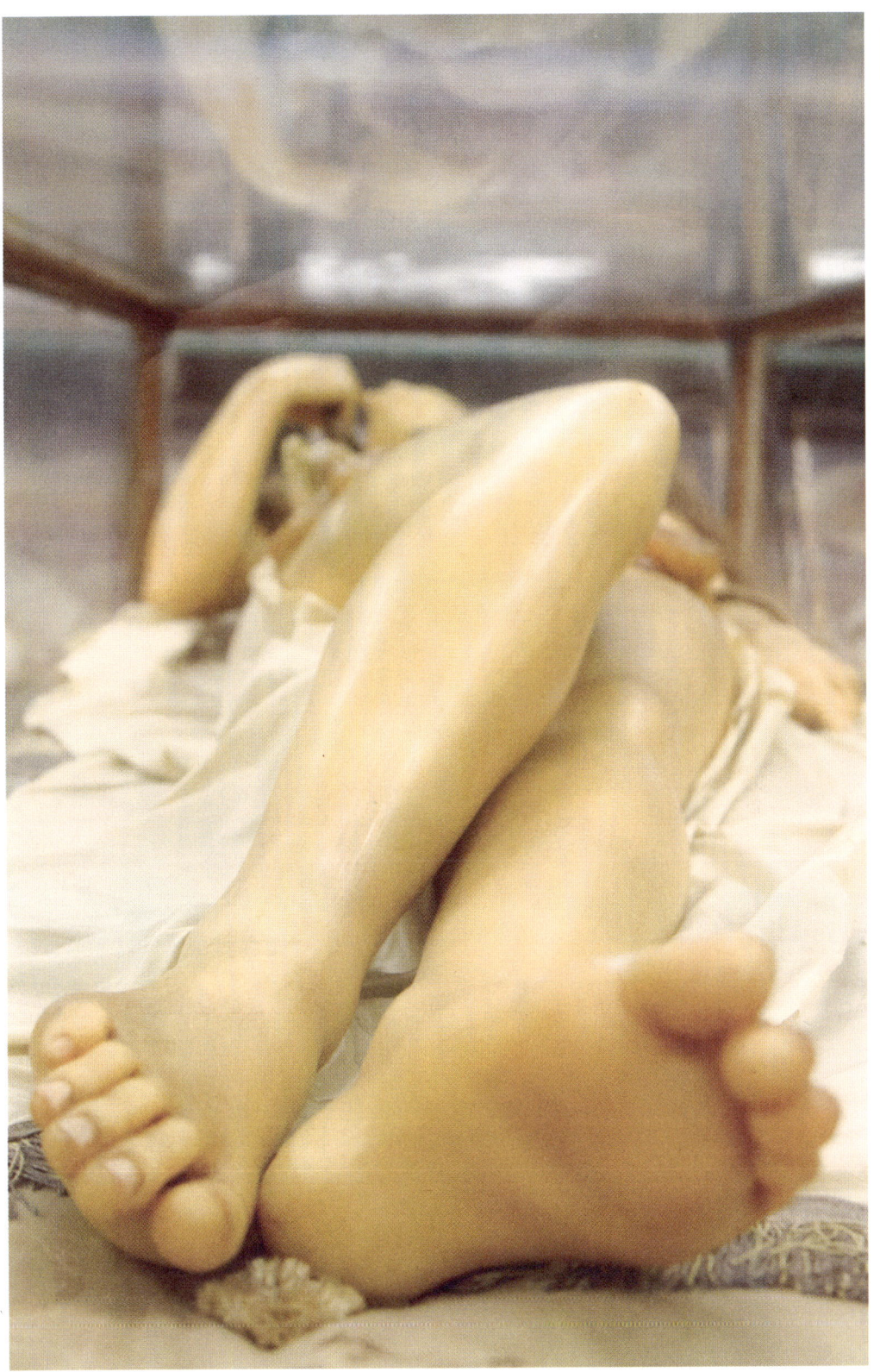

approaches, including what we would today define as science, aesthetics, and metaphysics, and which attempted to understand a divinely created natural world. The *Wunderkammer*, an expression of this worldview, was organized not by scientific principles as we now understand them, but as a microcosm of the universe intended to encourage the beholder to marvel at the wonder of God's works. At this time of unprecedented global exploration, marvels abounded, as European voyages of discovery returned with shiploads of new species, artefacts, and stories from previously unimagined or little-known civilizations.

Leopold II appointed his court physician and natural philosopher Felice Fontana (1730–1805) to oversee the creation of his new museum; he remained director of La Specola until his death. Inspired by the famous anatomical wax museum in Bologna that had been established some thirty years earlier, Fontana

fig. 10

fig. 11

fig. 10 The perfect proportions of the Venus de' Medici, a marble sculpture created in the first century BCE, were essential viewing on the Grand Tour. In the eighteenth century, she still bore remains of red on her lips and gold-leaf on her hair. These were rubbed away during restoration in 1815.

fig. 11 Sandro Botticelli's The Birth of Venus (1482–85), commissioned by the Medici family, was a must-see on the Grand Tour.

employed one of its sculptors, Giuseppe Ferrini (d. 1815), as artistic leader of his own dedicated in-house workshop for the creation of wax models. Fontana's grand and ambitious aim was to create an encyclopaedia of the human body in wax—one that would render human anatomy accessible and understandable to the general public, instructing people in scientific principles and the divine architecture of the human body. Fontana hoped that it would end the need for cadavers in the teaching of anatomy:

> *If we succeeded in reproducing in wax all the marvels of our animal machine, we would no longer need to conduct dissections, and students, physicians, surgeons, and artists would be able to find their desired models in a permanent, odour-free, and incorruptible state.*

The wax models produced by the workshop at La Specola were posed as if alive, healthy, and pain-free, in an attempt to distance the study of anatomy from the contemplation of death and bloody internal organs. The Anatomical Venus was further removed from notions of death and the corpse as she drew on the historical and artistic figure of the Roman goddess of love, beauty, and fertility, evoking a long history of paintings and sculptures of placid, idealized nudes. In fact, each pristine wax model at the museum was the product of the careful study of cadavers that were delivered from the nearby Santa Maria Nuova hospital. Although the collection fell short of its ambition to render all further human dissection unnecessary, today, over two hundred years after their creation, La Specola's waxworks are still considered remarkably accurate, some of them demonstrating anatomical structures that had yet to be named or described at the time of their making.

The best known of all the wax artists, or ceroplasticians, employed at La Specola's workshop was Clemente Susini, who had trained as an artist and went on to teach at Florence's Academy of Fine Arts. Susini began as Ferrini's assistant

fig. 12

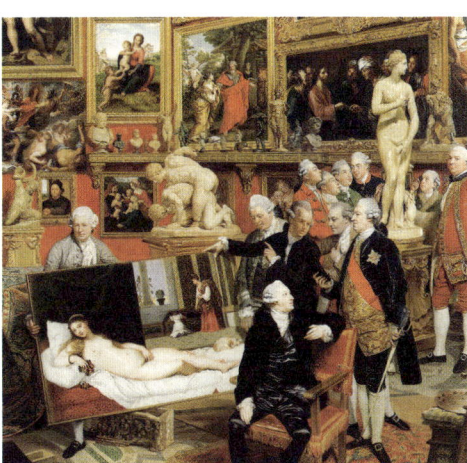

fig. 13

fig. 12 Titian's Venus of Urbino (1538) was commissioned by the Duke of Urbino as a gift for his young wife, Giulia Varano; it represents an allegory of marriage, and wifely duties of eroticism, fidelity, and motherhood.

in 1773, was promoted to lead modeller in 1782, and worked at the museum until his death, when he was succeeded by his assistant. It was under Susini's oversight that the museum created its finest and most iconic works, among them the ingenious, dissectible Medici Venus.

The figure of Venus was no random allusion; by the eighteenth century she had a long-standing relationship with Florence. As Rebecca Messbarger explains, the city had been nicknamed 'Venus-Fiorenza' in the sixteenth century to symbolize the fertility, happiness, and beauty of Florence under Medici rule. Depictions of Venus were also, Messbarger notes, a 'principal organizing theme' on the Grand Tour, a secular pilgrimage made by scores of wealthy and well-educated young European men, and occasionally women, to acquaint themselves with the artistic achievements of continental Europe. Italy's treasures loomed large in the Grand Tour, with Florence and its Venuses high on the list: Grand Tourists would visit Botticelli's iconic *Birth of Venus* (1482–85) in the Medici country villa, then stop off at the royal gallery of the Uffizi to pay homage to Titian's sensual *Venus of Urbino* (1538) and the *Venus de' Medici*, a Hellenistic sculpture and exemplar of perfect female proportions created in the first century BCE.

fig. 13 Detail of Johann Joseph Zoffany's The Tribuna of the Uffizi (1772–77), depicting Grand Tourists paying homage to the much-lauded artworks of the Uffizi Gallery. Titian's Venus of Urbino and the Venus de' Medici are prominently displayed.

Frontispiece of Wondertooneel de Natuur, Tome I (1706) by Andries van Buysen (1698–1747), depicting the Wunderkammer of wealthy Amsterdam textile designer and merchant Levinus Vincent (1658–1727).

The Medici Venus was, in part, an attempt to add La Specola to the list of 'must-sees' on the Italian leg of the Grand Tour. By creating a Venus that was educational as well as beautiful, Leopold II was making a statement about his values in contrast to those of the Medici, which he saw as frivolous and decadent.

Of all the waxworks in La Specola's collection, only the model that later came to be known as the Medici Venus—a play on the plural of *medico*, Italian for doctor, as well as an allusion to the *Venus de' Medici*—was 'demountable', or dissectible. As such, she was not only an object of art and anatomy, but also a *Wunderkammer*-worthy marvel, a 'novel Instrument, an unexpected physics machine' (Messbarger, 2010). Other full-length Anatomical Venuses were made for La Specola to illustrate the female body frozen in different stages of dissection. Often referred to as 'Slashed Beauties' or 'Dissected Graces', they each demonstrate different internal systems of the body.

fig. 14 Nineteenth-century engraving of La Specola anatomical museum, named after the building's observatory (specola) as *seen from Boboli Gardens, Florence, Italy.*

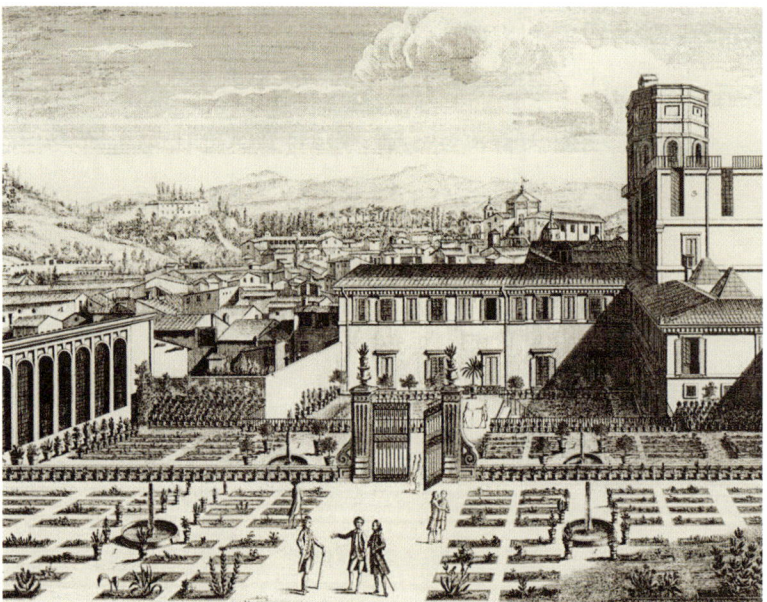

fig. 14

fig. 15

fig. 15 A cabinet of curiosity housing a mix of naturalia *and* artificialia, painted in 1675 by Domenico Remps (1620–99).

La Specola opened its doors to the public on 21 February 1775. The museum presented a microcosm of creation, with exhibits ranging from plants and minerals to taxidermic animals and scientific instruments. At the heart of the collection, and by far its most popular offering, was Felice Fontana's encyclopaedia of the human body in wax; added to over the years, today it comprises approximately 1,400 pieces, eighteen of which are whole, life-sized figures (including skinless male models). Each wax piece was accompanied by a colour illustration that pointed out important structures. Thus, the wax figures were an attempt to provide the general public with everything they needed to understand the human body—the pinnacle of God's creation—in an intuitive and pleasurable way, without need of lecturer or text. By consolidating private collections with customized wax models under one roof and making them available free of charge—albeit with separate opening hours for the lower classes, 'provided they were cleanly clothed'—Leopold II defined his own 'enlightened absolutist' idealism. His philosophy was in stark contrast to that of his sister, Marie Antoinette (1755–93), who had married the French king Louis XVI (1754–93). In France, not long after La Specola's Medici Venus was crafted, Marie Tussaud (1761–1850) was making wax models of the most famous heads rolling from the guillotine during the French Revolution.

From 1771 to 1893, Fontana's wax workshop produced more than 2,500 wax models for La Specola and a variety of other museums, including life-sized bodies and small anatomical details. A full set of 1,192 wax models was commissioned by Leopold II's brother, Emperor Joseph II (1741–90), and transported

fig. 16 Engraving by Giuseppe Zocchi of the hospital of Santa Maria Nuova, Florence (1744). The bodies that served as models for La Specola's waxes were sourced here.

fig. 16

over the Alps to Vienna on the backs of mules and labourers. Intended for use in training military surgeons at the Josephinum, the medico-surgical academy founded by Joseph II in 1785, the models were dismissed by some as expensive, frivolous toys despite their anatomical accuracy, perhaps because the Viennese middle class trusted neither aristocratic opulence nor popular pleasures. Napoleon Bonaparte (1769–1821) also ordered forty cases, which unfortunately never arrived in Paris, ending up instead in Montpellier, France, where they can still be seen at the Faculty of Medicine's anatomical museum. Today, Anatomical Venuses created at La Specola's workshop can be viewed at museums in Budapest, Pavia, Bologna, and London. A large collection of exquisite waxes still on display

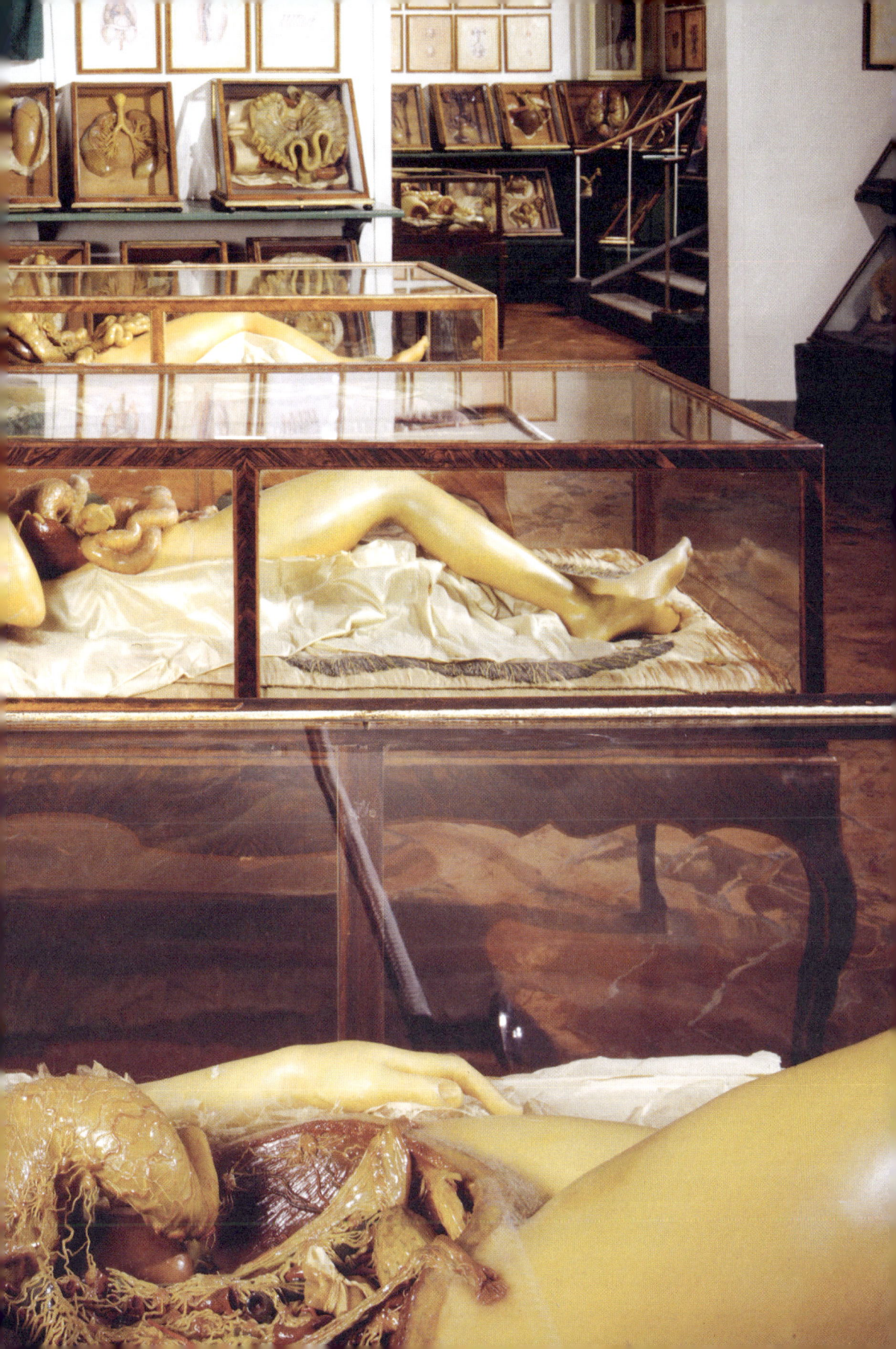

PREVIOUS
An interior view at La Specola showing three Anatomical Venuses surrounded by smaller waxes and illustrations of anatomical details.

in Cagliari, Sardinia, Italy, were made late in life by Clemente Susini, without Felice Fontana's supervision. Fontana notoriously gave his artists very little freedom, and the Cagliari waxes are considered among Susini's finest works.

The virtuosity of Susini's Medici Venus called not only upon the conventions of fine art, but also upon a considerable tradition of dissectible female figures. Anatomical illustrations had often taken the form of 'fugitive sheets'—in which paper flaps could be pulled back or moved to reveal the structures beneath—as well as static, cutaway views in which internal organs were made visible to simulate an imaginary dissection. The Medici Venus also evoked religious precedents, most notably statues of the *Mater gravida* or 'our lady of expectations'—in which a pregnant Mary was shown with the baby Jesus visible inside the womb

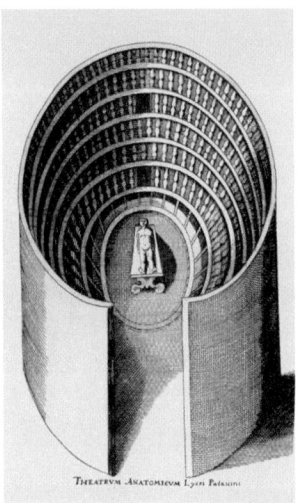

fig. 17

fig. 18

fig. 19

fig. 17 Seventeenth-century engraving of the anatomical theatre, University of Padua, built in 1594.

fig. 18 A public dissection in Padua's anatomical theatre. Frontispiece from Johann Vesling's Syntagma Anatomicum (1647).

through a door or cutaway. Also significant is a wooden, demountable, pregnant anatomical Eve from the seventeenth century, whose internal organs and fetus come into view when her breastplate is removed, her genitals discreetly hidden by a wreath of carved leaves. Felice Fontana eventually came to believe that wood was a better medium than wax for creating anatomical teaching models, as it was less fragile and the students could demount and reassemble the model, thus intuitively learning the relationship between the internal structures. He spent the last years of his life working on a prototype anatomical male model of painted wood, dissectible into 3,000 pieces. Due to fluctuations in humidity that changed the size of the pieces, it was never realized.

The most common precursors of the Medici Venus were smaller dissectible female figures (or, less often, male figures) crafted of ivory, wood, and other media, known as 'anatomical manikins' (*see also* pages 52–53). The majority of manikins were crafted in Germany in the seventeenth and eighteenth centuries. Each model is about the size of a hand and reclines on its own little bed, often on a pillow of cloth or ivory. Their organs, though dissectible, lack detail and accuracy, and are suggestive rather than descriptive. Female figures contain, in addition to their other rough-hewn organs, a tiny fetus, sometimes attached to the body by a red silk thread. These enigmatic and seductive toys may have been tools for teaching expectant mothers or midwives about childbirth. Or they might, as many scholars believe, have been more decorative in nature, intended as collectables for *Wunderkammern*, or as a way for physicians to advertise their professional standing.

The first full-sized female instructional anatomical wax models began to be created in the early eighteenth century. In 1719, French surgeon and anatomist Guillaume Desnoues (1650–1735) (*see also* pages 96, 97, 100–101) publicly exhibited a dissectible wax woman featuring a newborn child with the umbilical cord

fig. 19 Engraving of Pieter Pauw, who instigated the building of Leiden's anatomical theatre, performing an anatomical dissection there (1615). The skeleton holds a banner reading 'Mors ultima linea rerum' ('Death is everything's final limit').

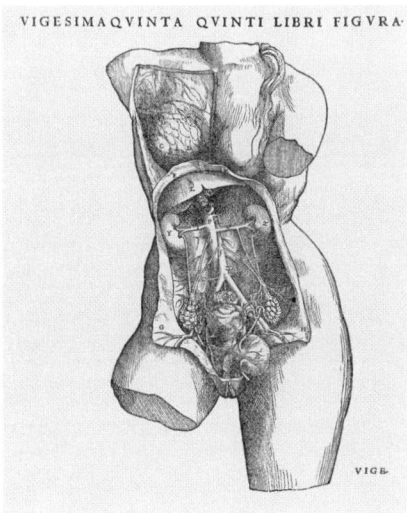
fig. 20

fig. 21

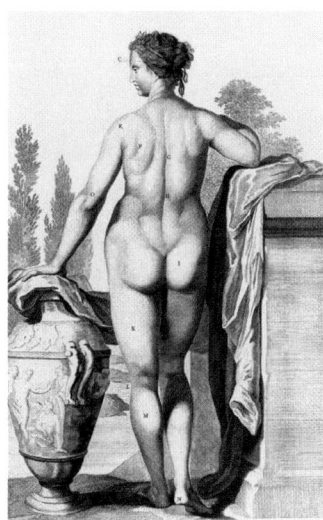
fig. 22

still attached. Little more than a decade later, in London, Paris-trained anatomist and modeller Abraham Chovet (1704–90) exhibited a female depicted as though in the painful process of being vivisected. The figure was represented as '...a woman big with child chained upon a table; supposed to be opened alive. In the face there is a lively display of the agonies of a dying person, the whole body heaving and the hands clinched, the action suitable to the character of the subject.' (*General Evening Post*, 1734). The circulation of her blood was demonstrated by a network of blown-glass tubes coursing with blood-red claret. In France, Marie-Catherine Biheron (1719–86), an artist with experience of the dissection room, had, from the age of sixteen, created wax anatomical models that she exhibited to the public on Wednesdays for three francs per person. Biheron later displayed her pieces in London, where they were reportedly admired by the famous Scottish surgeons John and William Hunter, and eventually purchased by Empress Catherine II (1729–96). In late-eighteenth-century Vienna, court sculptor Josef Müller-Deym (1752–1804) displayed a dissectible female wax model of a pregnant woman. Müller-Deym also made other kinds of wax ladies:

fig. 20 Detail of the female reproductive system, from Andreas Vesalius's De Humani Corporis Fabrica (1543).

figs 21, 22 Idealized female nudes, from Bernhardus Siegfried Albinus's Tables of the Skeleton and Muscles of the Human Body (1749).

OVERLEAF
Hand-coloured woodcut frontispiece of Andreas Vesalius's De Humani Corporis Fabrica, Libri Septem (1543).

ANATOMY IS AN IMPORTANT PART OF NATURAL PHILOSOPHY; SINCE IT EMBRACES THE STUDY OF MAN AND MUST PROPERLY BE REGARDED AS THE PRIME FOUNDATION OF THE WHOLE ART OF MEDICINE AND THE SOURCE OF EVERYTHING THAT CONSTITUTES IT.

Excerpt from De Humani Corporis Fabrica, Libri Septem by Andreas Vesalius (1543).

Andreas Vesalius

Bruxellensis

Invictissimi Caroli V. Imperatoris Medicus.

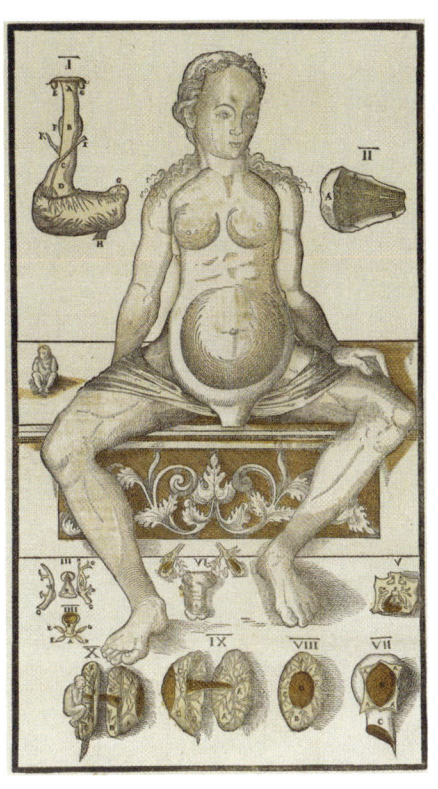
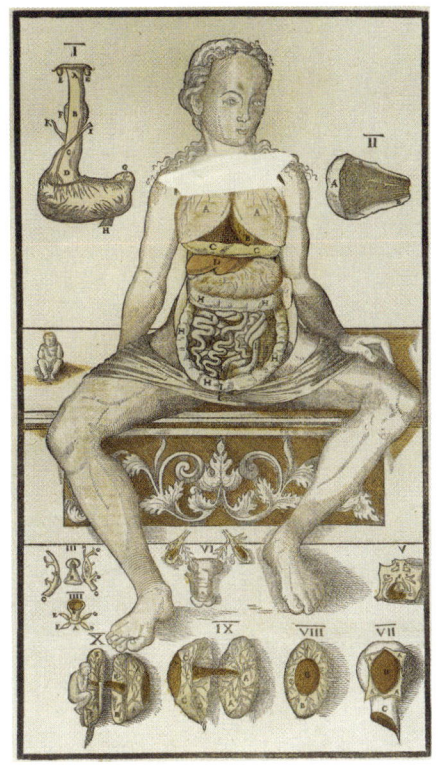
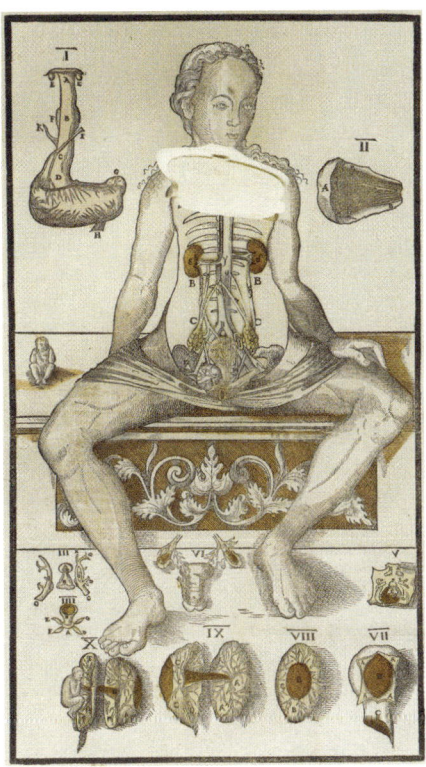
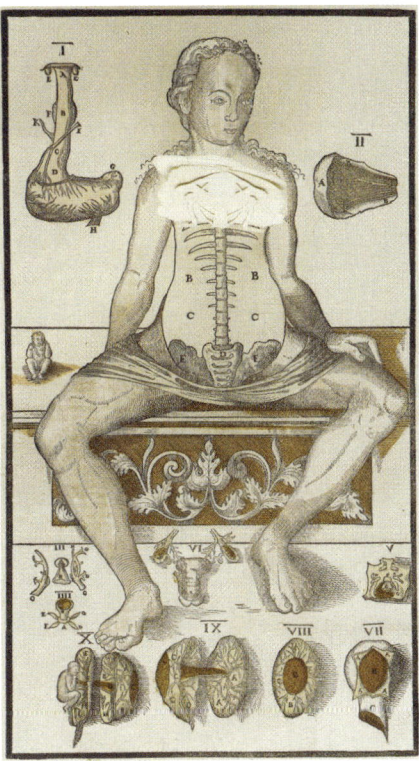

Anatomical 'fugitive sheets' (1573): drawings with paper flaps that can be lifted in successive layers to reveal the bones and viscera beneath.

in 1791, police invaded his cabinet and destroyed a number of models intended for private collectors, 'the sight of which', according to Joseph Richter's satirical novel *Die Eipeldauer-Briefe* (1785–1797) 'would in most Christian contemporaries have overturned the good teachings of their preachers'.

Human anatomy had first become of great interest to artists, natural philosophers, and the general public during the Renaissance. It was particularly

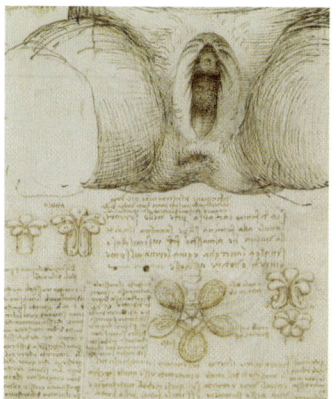
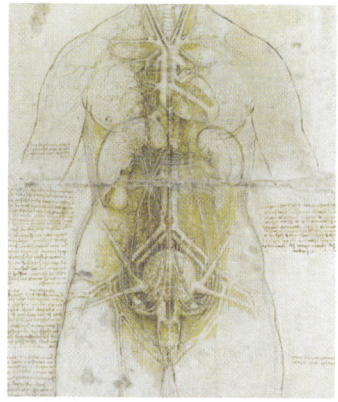
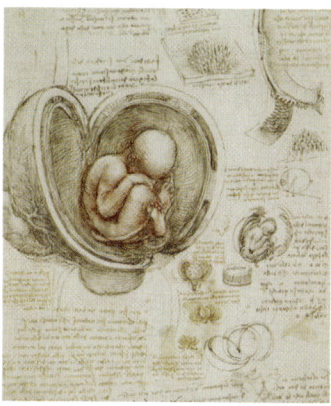
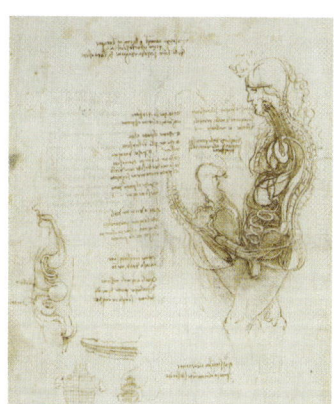
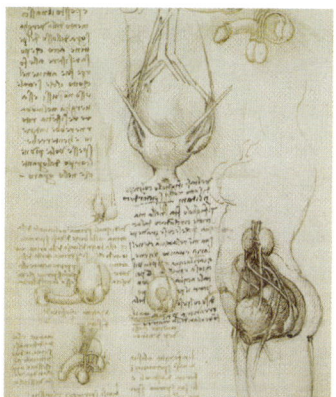
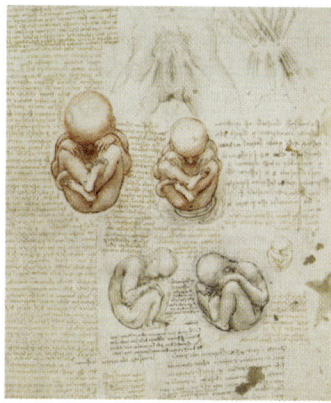

fig. 23

fig. 23 Leonardo da Vinci's anatomical sketches in pen and ink over chalk (1490–c. 1511). From top left to bottom right: vulva and anus; cardiovascular system and principal organs of a woman; fetus in the womb; the act of coitus; male and female reproductive systems; a fetus, and the muscles attached to the pelvis.

important to visual artists to understand the underlying structures of the body in order to depict human subjects more realistically. To this end, artists conducted their own dissections—more, some say, than the anatomists of the era. Artists also often used wax in their representations, especially in the crafting of three-dimensional écorchés—skinless figures demonstrating musculature—as studies for finished works. Among the best-known écorchés is Ludovico Cardi's, known as *La bella notomia* (The Beautiful Anatomy) or *Lo scorticato* (The Flayed), made around 1600 in Florence. Its immense popularity has led to numerous casts being made from his prototype in bronze and plaster ever since. Andrea del Verrocchio (1435–88) of Florence was the first known artist to make practical use of wax écorchés for study in art schools.

Leonardo da Vinci (1452–1519)—del Verrocchio's most famous student—is said to have dissected more than one hundred bodies, and famously 'sketched cadavers he had dissected with his own hand' (Vasari, 1991). Da Vinci had ambitions to publish a 120-volume set of his anatomical works in conjunction with Marc' Antonio della Torre (1481–1511), Professor of Anatomy at Padua and Pavia Universities, but the project was abandoned after the latter's untimely death. Michelangelo Buonarroti (1475–1564), Da Vinci's younger contemporary, studied anatomy in Florence for twelve years, and is said to have accepted a commission for the Church of the Holy Ghost on the condition that he was paid in cadavers.

fig. 24 Écorché (a study of the human muscles without skin) in pen and ink, by Ludovico Cardi (1559–1613), known as Il Cigoli.

fig. 25 Michelangelo Buonarotti's preparatory sketch of the Libyan Sibyl for the Sistine chapel (1510–11).

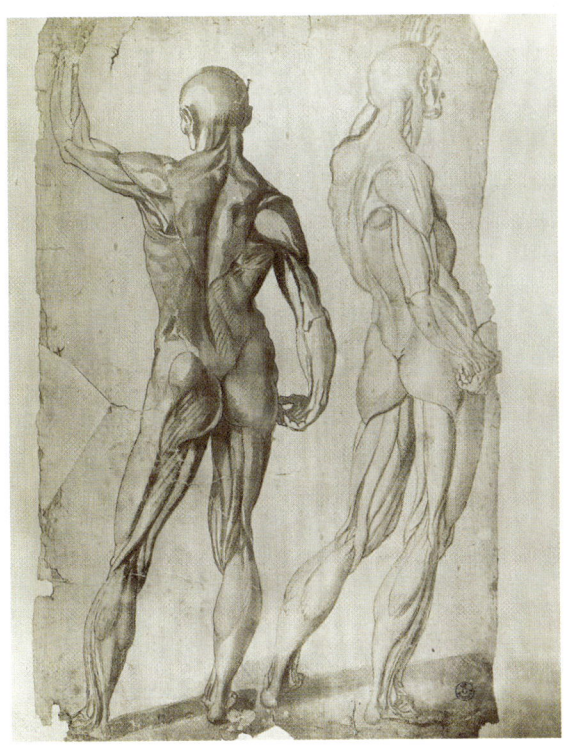

fig. 24

fig. 25

It was Flemish professor of surgery and anatomy at Padua University, Andreas Vesalius (1514–64) who revolutionized the study of anatomy at this time. His monumental work, *De Humani Corporis Fabrica* (On the Fabric of the Human Body) was lavishly illustrated with woodcuts thought to be by Titian's studio in Venice. This masterwork was published in 1543—the same year as Copernicus's controversial publication *De Revolutionibus Orbium Coelestium* (On the Revolutions of the Heavenly Spheres), which described the planets' rotation around the Sun and debunked the canonical belief that the entire cosmos revolved around Earth. Both books marked dramatic paradigm shifts in their respective disciplines, and questioned beliefs that had been founded on faith rather than empirical observation.

OVERLEAF
L'Ange Anatomique (The Flayed Angel) (1746), a mezzotint of a beautiful, fashionably coiffed anatomized woman, by Jacques-Fabien Gautier d'Agoty, a pioneer of medical colour mezzotint printing.

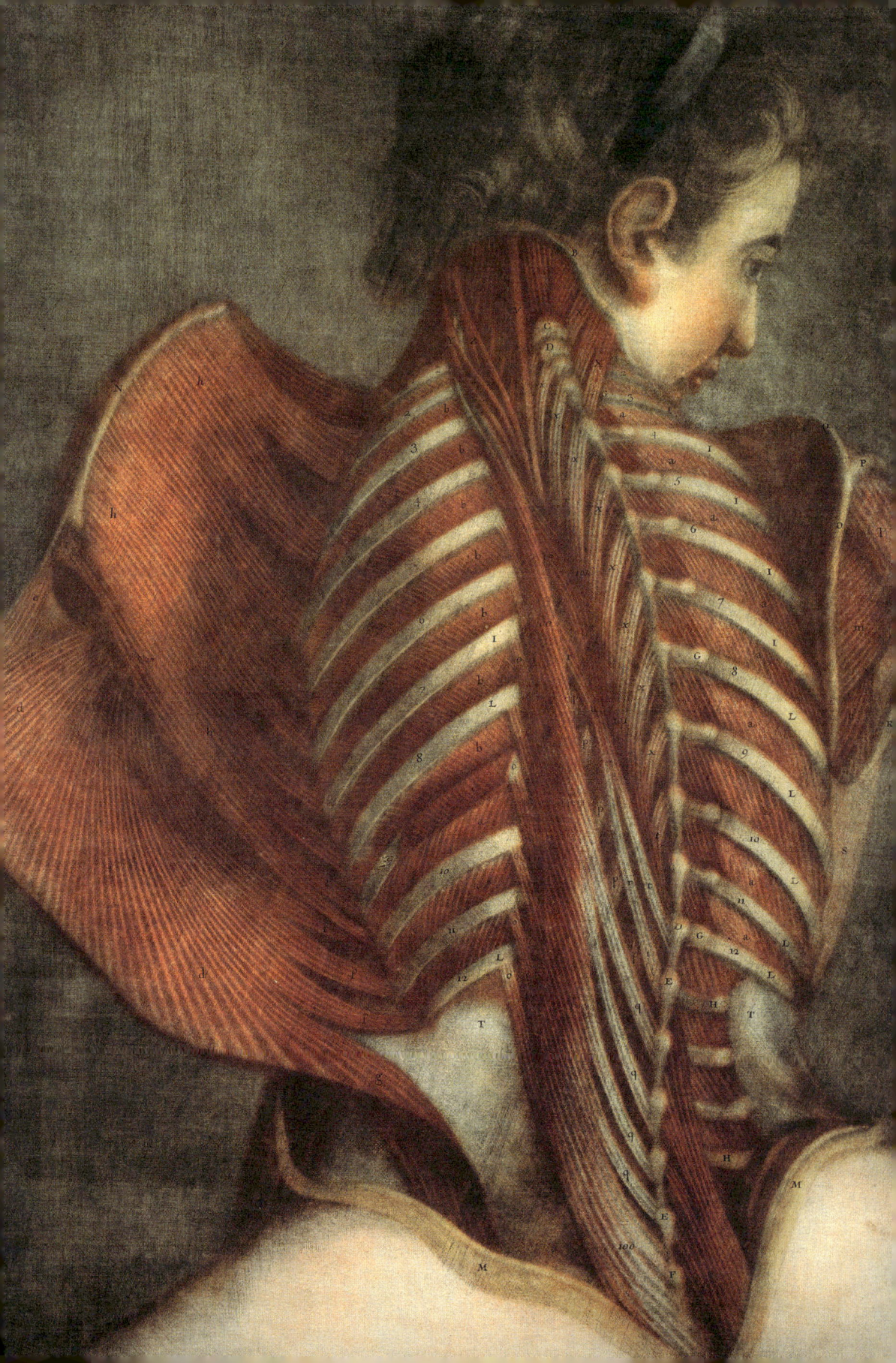

FOR·MEN TO·BE INSTRUCTED, THEY·MUST·BE SEDUCED·BY AESTHETICS, BUT·HOW CAN·ANYONE RENDER THE·IMAGE OF·DEATH AGREEABLE?

Arnaud-Eloi Gautier d'Agoty (1741–80), son of Jacques-Fabien Gautier d'Agoty, describes the challenge faced by anatomical illustrators.

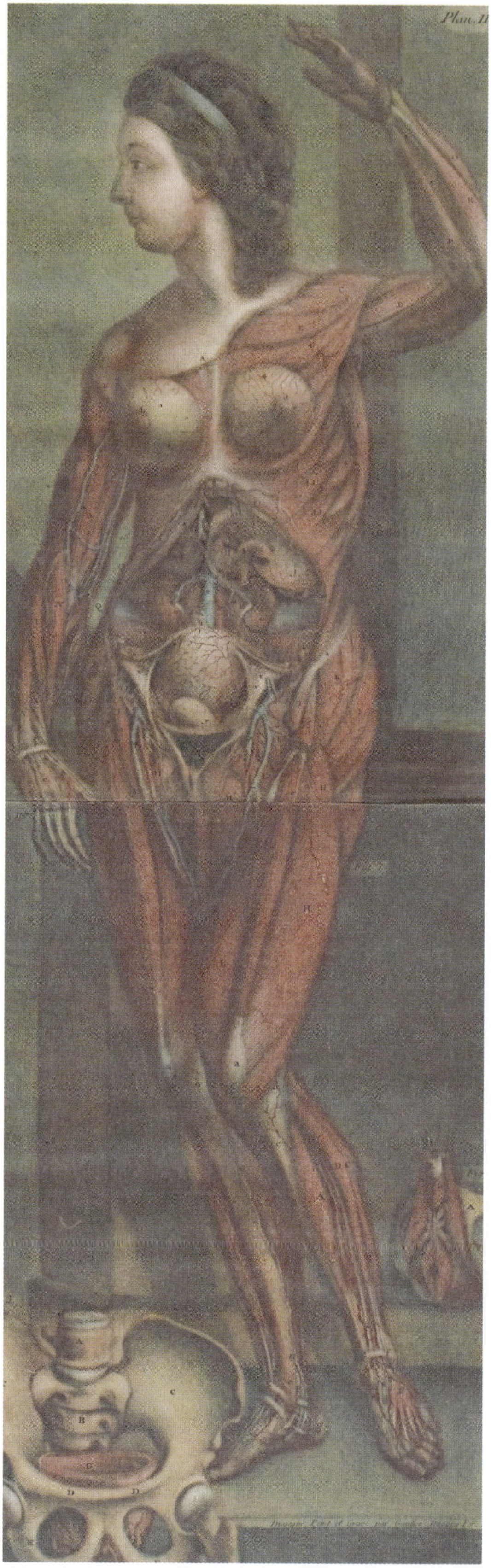

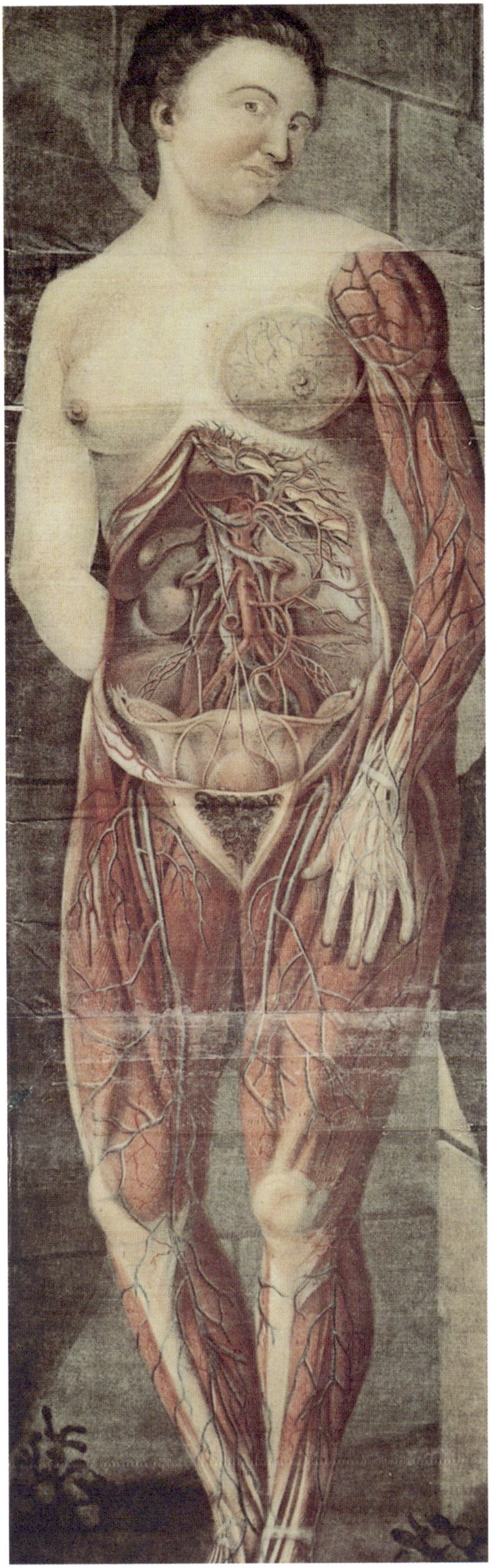

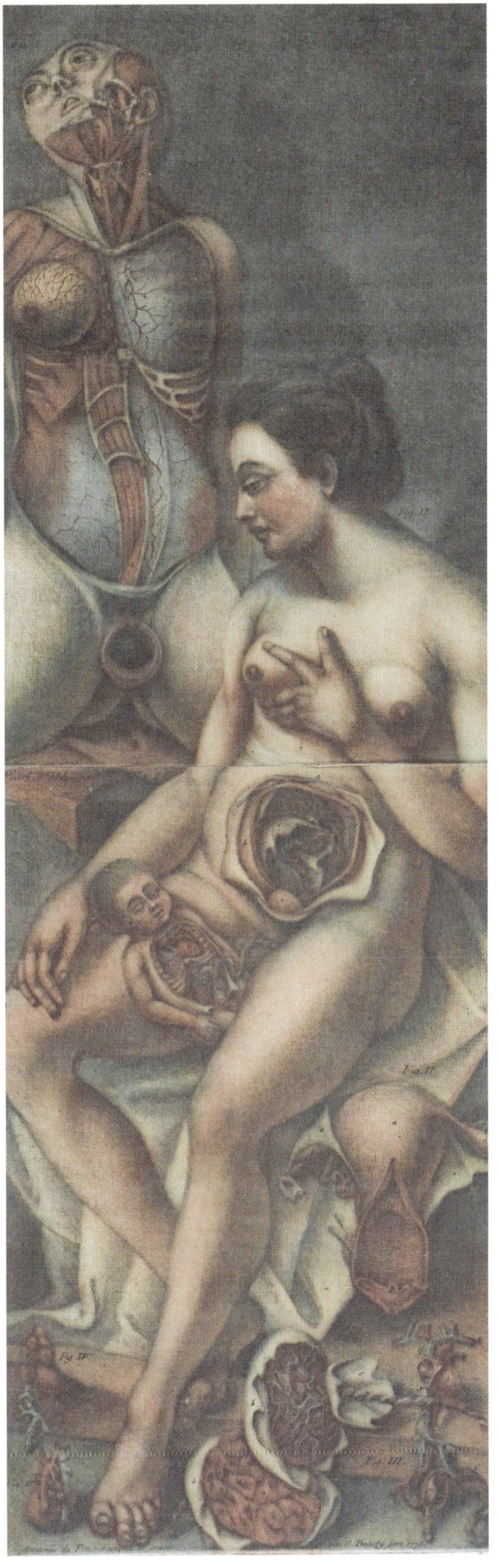

PREVIOUS

Elegantly posed écorchés showing the female reproductive system, by Jacques-Fabien Gautier d'Agoty. Taken from Anatomie des parties de la génération de l'homme et de la femme *(1773) (left page, both; right page, right side) and* Anatomie generale des visceres en situation *(1772) (right page, left side).*

Vesalius, unlike many contemporary anatomists, conducted his own dissections of cadavers, which led him to recognize a variety of mistakes in the work of Claudius Galen (*c.* 129–216 CE), on whose *Hippocratic Corpus*, an anthology of texts about the body from the fourth and fifth centuries BCE, much anatomical knowledge of the time relied. The errors, Vesalius discovered, stemmed from the fact that the old master performed most of his dissections on animals (especially pigs, dogs, and apes) rather than humans, due to prohibitions about human dissection in Roman law. His inaccuracies had been handed down for generations due, in part, to the way in which dissections were conducted, with the professor standing above the proceedings and reading aloud from Galen, while a labourer conducted the dirty work of dissection.

Vesalius understood the power of illustration; his woodcuts, he explained, were not 'executed merely as simple outlines, like ordinary diagrams in textbooks, but have been given a particular pictorial quality'. *De Humani Corporis*

fig. 26

figs 26, 27 Dissectible, pregnant anatomical figure and removable pieces, carved to scale from linden wood (c. 1700).

Fabrica was read by unprecedented numbers of people, due to new printing technology that enabled the mass production of high-quality, large-scale illustrations. In *Fabrica*'s wake, dozens of artistic and expressive anatomical atlases—many of them large, luxurious affairs, of interest to private collectors as much as students of anatomy—were produced. Some of the most memorable are the work of French artist and anatomist Jacques-Fabien Gautier d'Agoty (1710–85); he was among the first to create an atlas in full colour, by a mezzotint process of his own devising. His images have a dreamy, painterly quality; indeed, he sometimes even varnished his colour prints in imitation of oil paintings.

They contrast with the annotated diagrammatic illustrations by Henry Carter (1831–97) that appeared in *Gray's Anatomy*, which was published in England in 1858. Avoiding the inclusion of any extraneous detail, Carter's style was designed for optimum clarity in pressurized clinical contexts and epitomizes the modern scientific method.

The modellers who created the anatomical wax models at La Specola and other wax workshops depended for the veracity of their work on the anatomical illustrations featured in anatomical atlases. Firstly, the modeller, usually taking advice from an anatomist or natural philosopher, would select an illustration, or illustrations, from trusted anatomical atlases by Vesalius, Albinus,

Haller, Mascagni, or another. Real human body parts would then be procured to work from, in order to ensure that all the individual parts were as accurate as possible. An Anatomical Venus was expensive and time-consuming to produce. Over two hundred cadavers were sometimes required to craft a single dissectible figure, owing to the speed with which bodies decayed, especially in the hot weather of the summer months.

The modeller would either take a cast of the prepared specimen or copy it by hand. As Susini's anatomical models are depicted with plump, rounded organs rather than deflated or even putrefying ones, British artist and anatomical ceroplastician Eleanor Crook believes that they were most often sculpted by hand from observation of a dissected cadaver, rather than directly cast from one. Once a model of inexpensive wax or clay had been approved, a plaster cast would be taken to serve as a mould, which could then be used repeatedly; many such moulds are still owned by La Specola today.

fig. 27

Next, the plaster moulds would be coated with soap or oil to ease release of the wax. The most commonly used waxes were beeswax; white 'Virgin wax' from Smyrna or Venice; or that of the Chinese scale insects *Ceroplastes ceriferus* and *Ericerus pela*, which produce a particularly fine, hard, white wax with a high melting point, well suited to modelling skin, though prohibitively expensive. The wax would be mixed with turpentine and other oils or fats to produce the required texture, as well as a mastic, or plant resin, to fortify and increase its stability, which was important for sustaining structure and retaining vivid colours. It would then be carefully heated and coloured with finely ground pigments, often highly precious or toxic, which had been sifted through cloth and dissolved in oil or turpentine. Thin layers of tinted wax would then be painted into, cooled and released from the mould. As most of the pieces were hollow, they were stuffed with rags or woodchips for support, although some, including Susini's Medici Venus, have metal frames. Hair was attached with varnish; eyelashes were individually implanted. Fine blood vessels and nerves were made of silk or linen fibres dipped in wax. The parts would be assembled, while attending to any flaws or damage. Finally, the model would be glazed in order to keep its surfaces free of dust and effect a realistic shine: another anatomical masterwork ready for display.

OVERLEAF
page 50
Early-seventeenth-century, life-sized wooden dissectible anatomical Eve, shown fully intact (left) and with her breastplate removed (right) to reveal viscera and the baby in her womb.

page 51
The same anatomical Eve with the top of her head lifted off to reveal the painted wooden brain beneath.

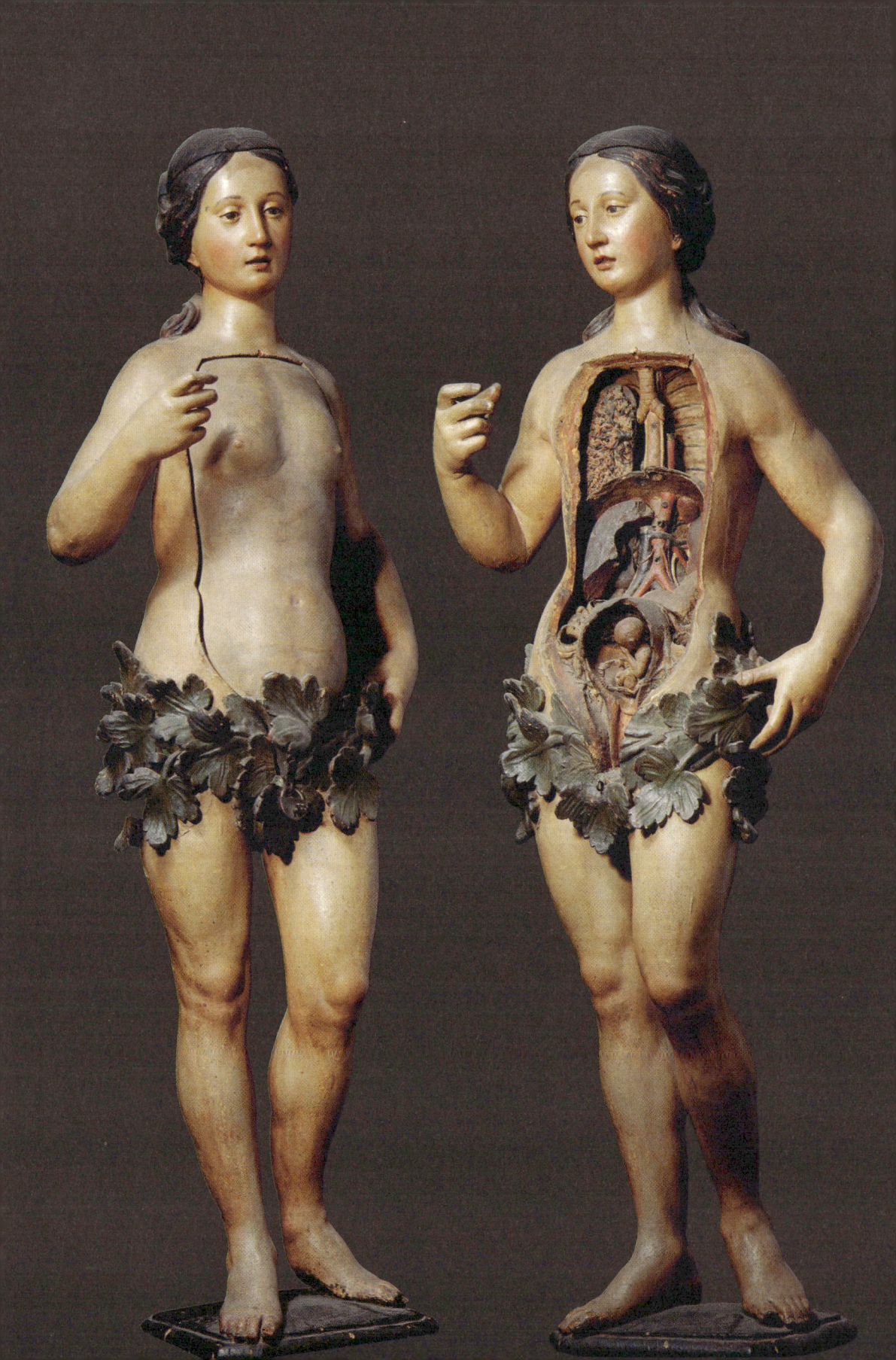

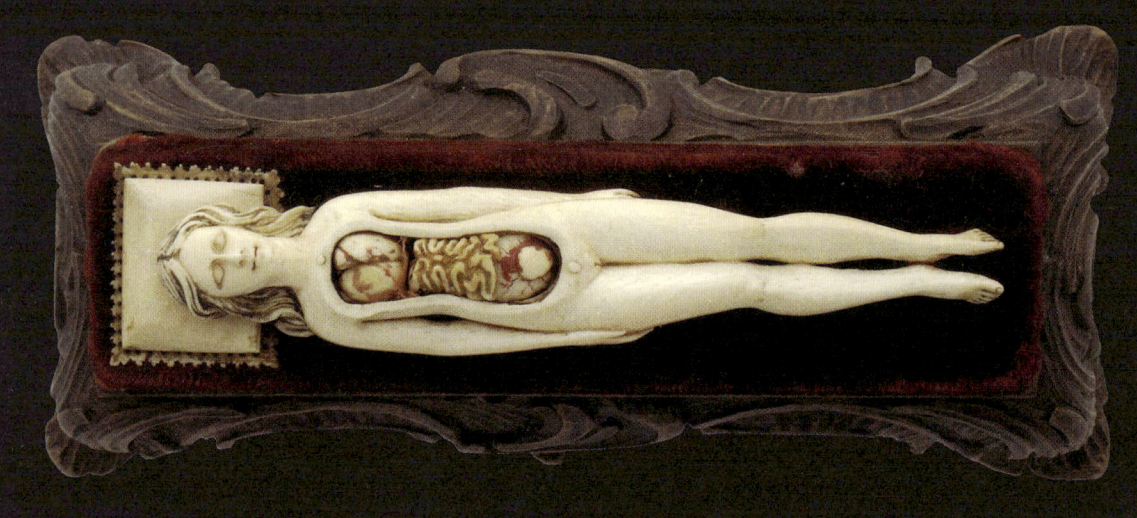

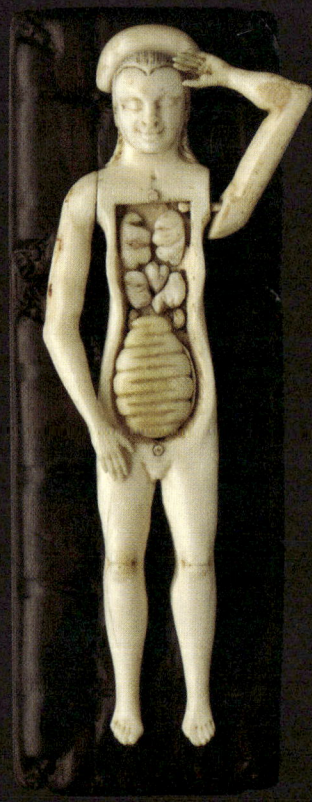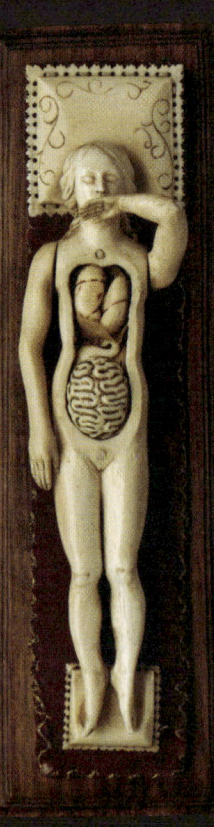

These miniature dissectible female ivory anatomical manikins were created in Europe during the seventeenth and eighteenth centuries. They were probably used by medical specialists and midwives to train their students, instruct young couples, or reassure pregnant women; the torso of each can be removed to reveal the internal organs. The following poem written by Italian obstetrician Joseph Fuardi de Fossau in 1786 accompanied one such manikin.

In Life's full bloom, when
labour's toil so near
My fellow sufferers' lot and
perils I do fear,
Come ye fair pupils, Lo,
I cast aside my shame
That Midwif'ries secrets may
reveal my frame.
Pierce it with keen enquiring
eye, and may
The child and mother's
nature then convey
New manifold devices to
your skilful art
That pining women may not
henceforth smart
Through cruel untaught
efforts, and not gasp
With their unborn in Death's
unpitying grasp.

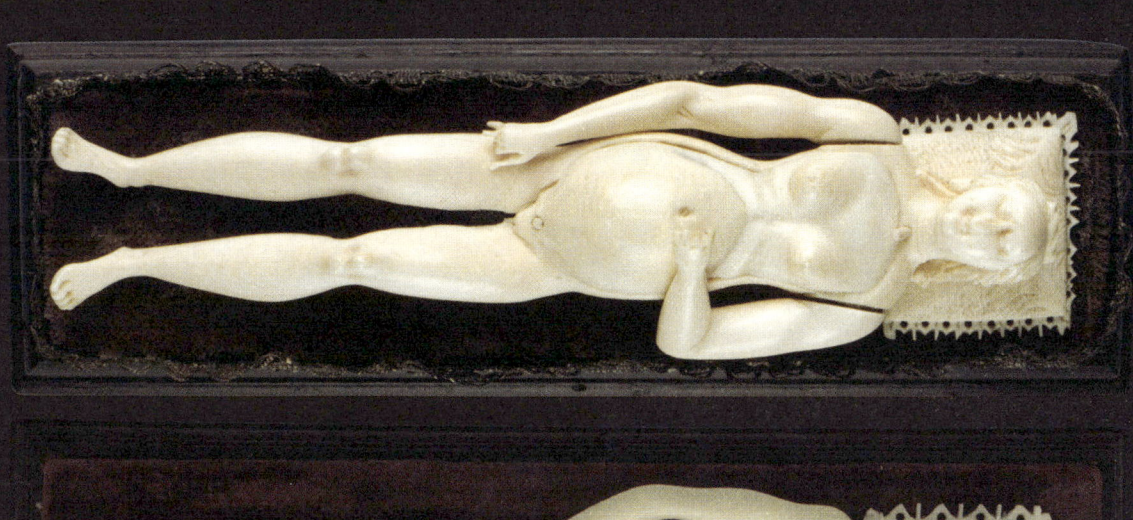
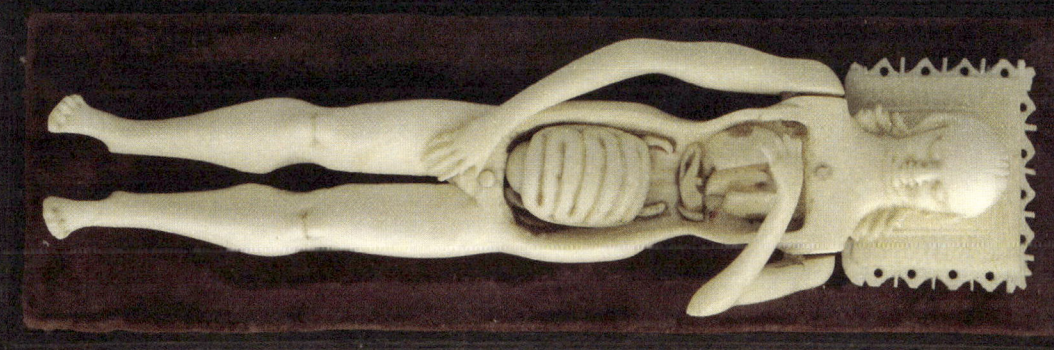

> IT IS NOT POSSIBLE TO CONTEMPLATE THE STRUCTURE OF THE HUMAN BODY WITHOUT FEELING CONVINCED OF SOME DIVINE POWER, DESPITE WHAT A FEW MISERABLE PHILOSOPHERS HAVE DARED TO SAY, IN M. FONTANA'S LABORATORY ONE KNEELS AND BELIEVES.

Louise Élisabeth Vigée Le Brun (1755–1842), official portraitist to Marie Antoinette, comments on La Specola in 1792.

Details from the meticulous anatomical waxes created by the workshop at La Specola. The finest blood vessels are made of silk or linen threads dipped in wax.

Anatomical model of the eye and lacrimal apparatus sculpted by wax modeller Cesare Bettini (c. 1850).

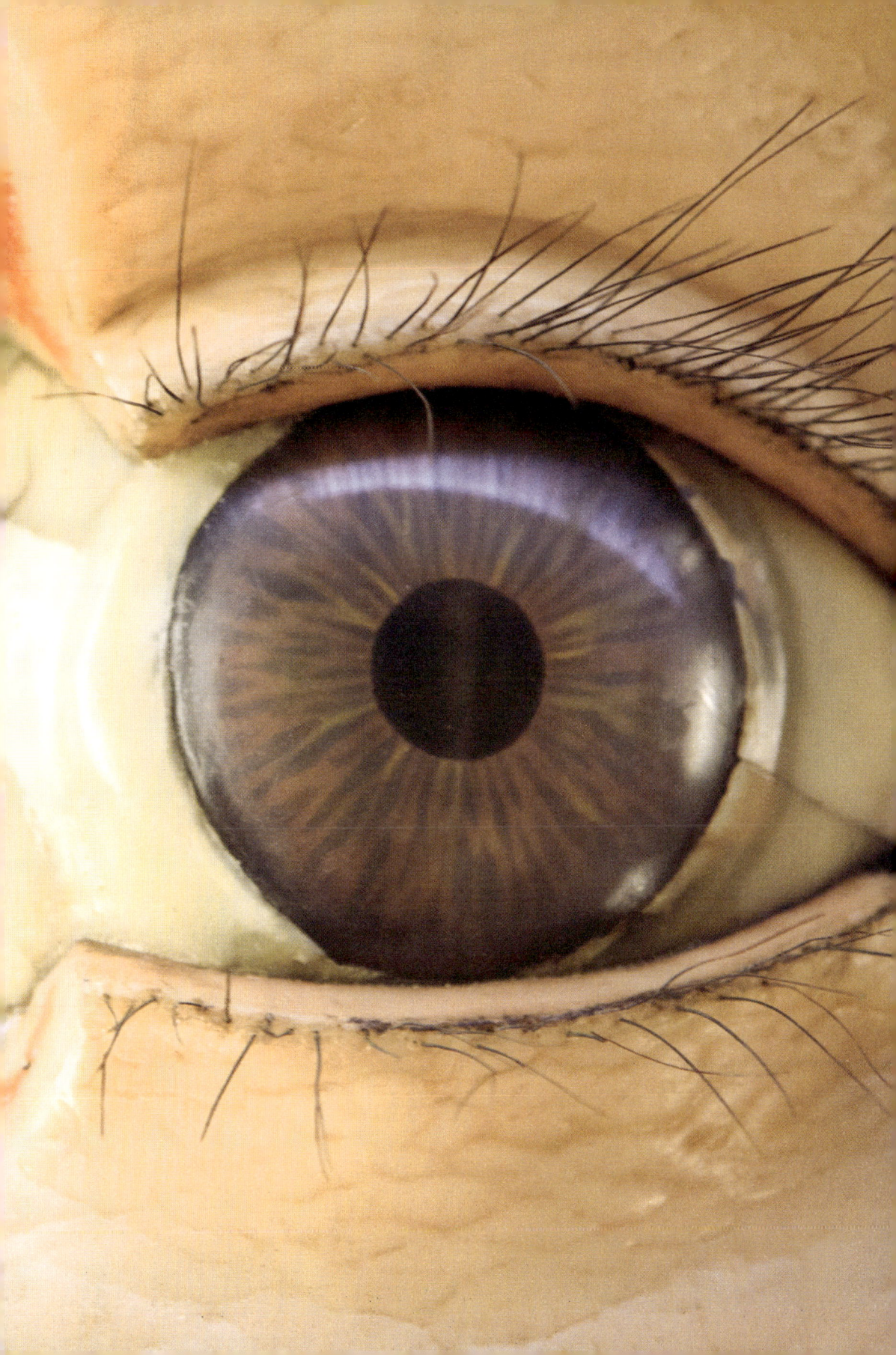

A parade of animal skeletons led by a life-sized écorché of a man, representing mankind's privileged place in the natural world, in the Muséum national d'histoire naturelle, Paris, France, established 1793.

TOP Madame Tussaud's waxworks of the heads of King Louis XVI and Marie Antoinette after they were guillotined, possibly cast from the original death masks. Photographed in London in the early 1960s.

ABOVE Late-nineteenth-century photograph of the atelier of wax modeller Emil Eduard Hammer in Munich, Germany. Hammer made models for the exhibitions of curiosities known as panopticons and for medical museums.

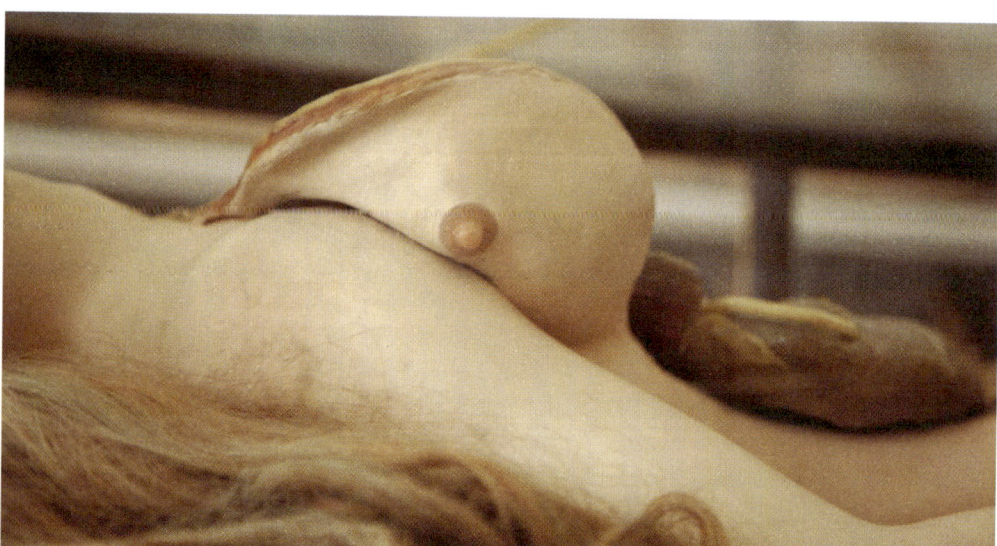

THIS PAGE & OPPOSITE *This Anatomical Venus, produced by the workshop at La Specola between 1784 and 1788, is displayed in her original rosewood and Venetian glass case at the Josephinum, Vienna, Austria.*

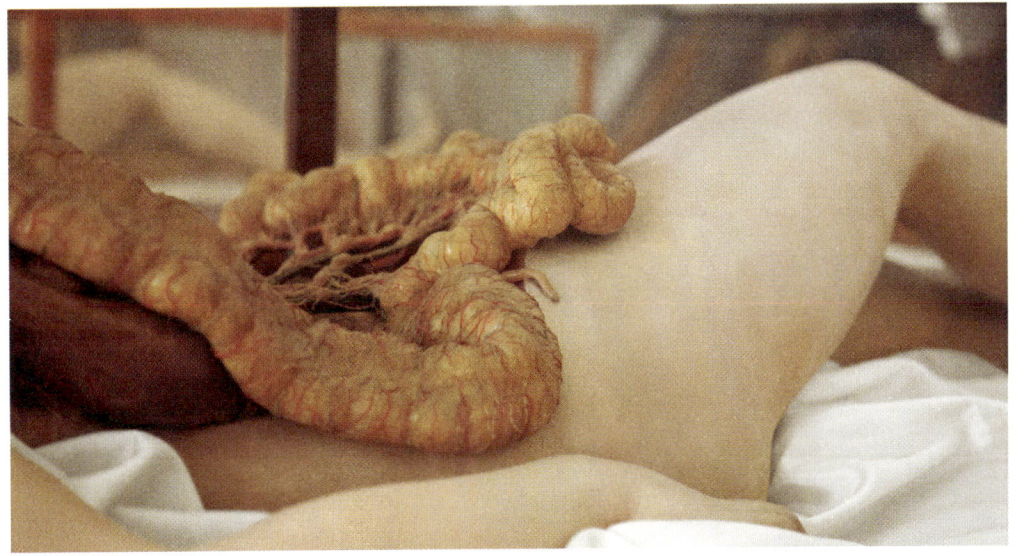
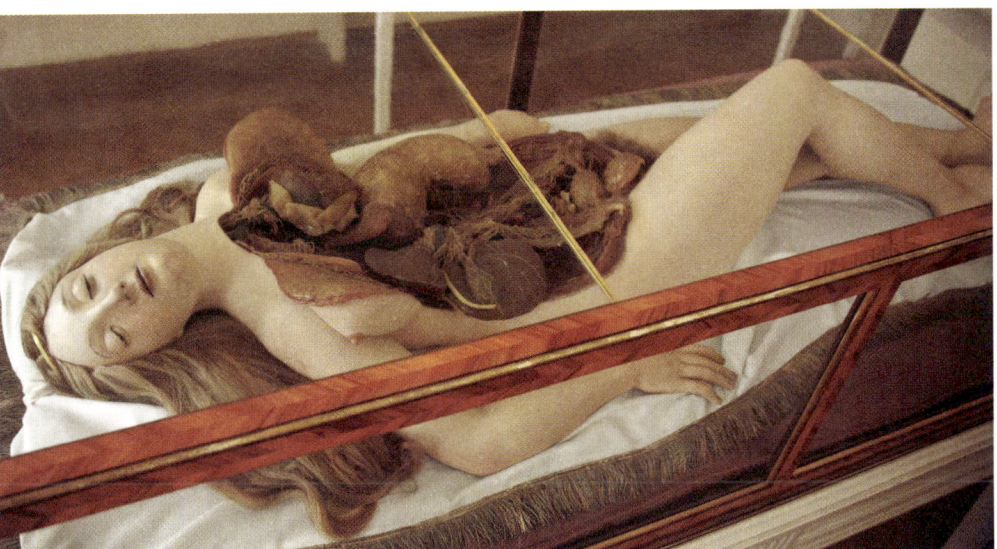
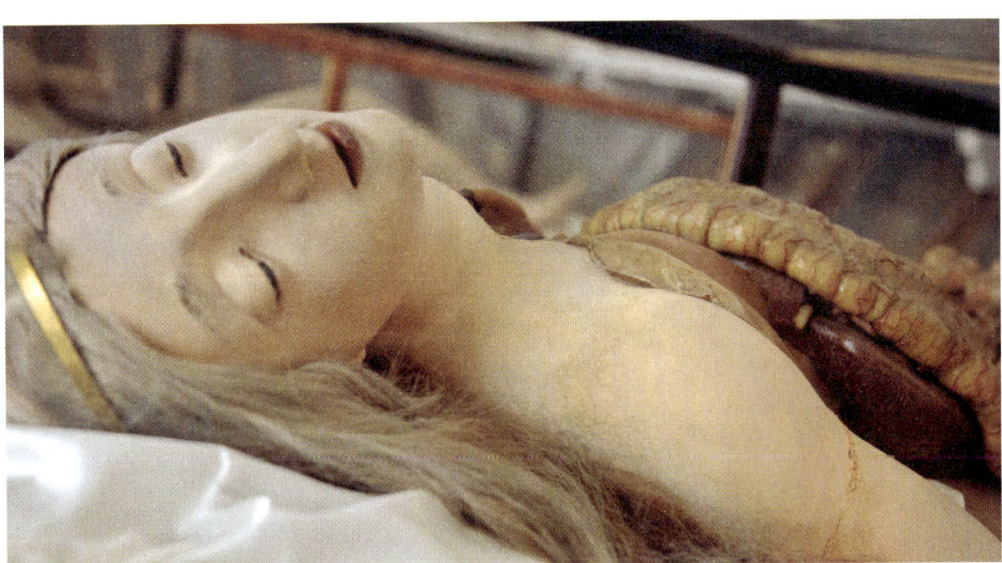

OVERLEAF *Half-scale wax Anatomical Venus in various states of dissection. She is likely to have been a study for the iconic Medici Venus at La Specola.*

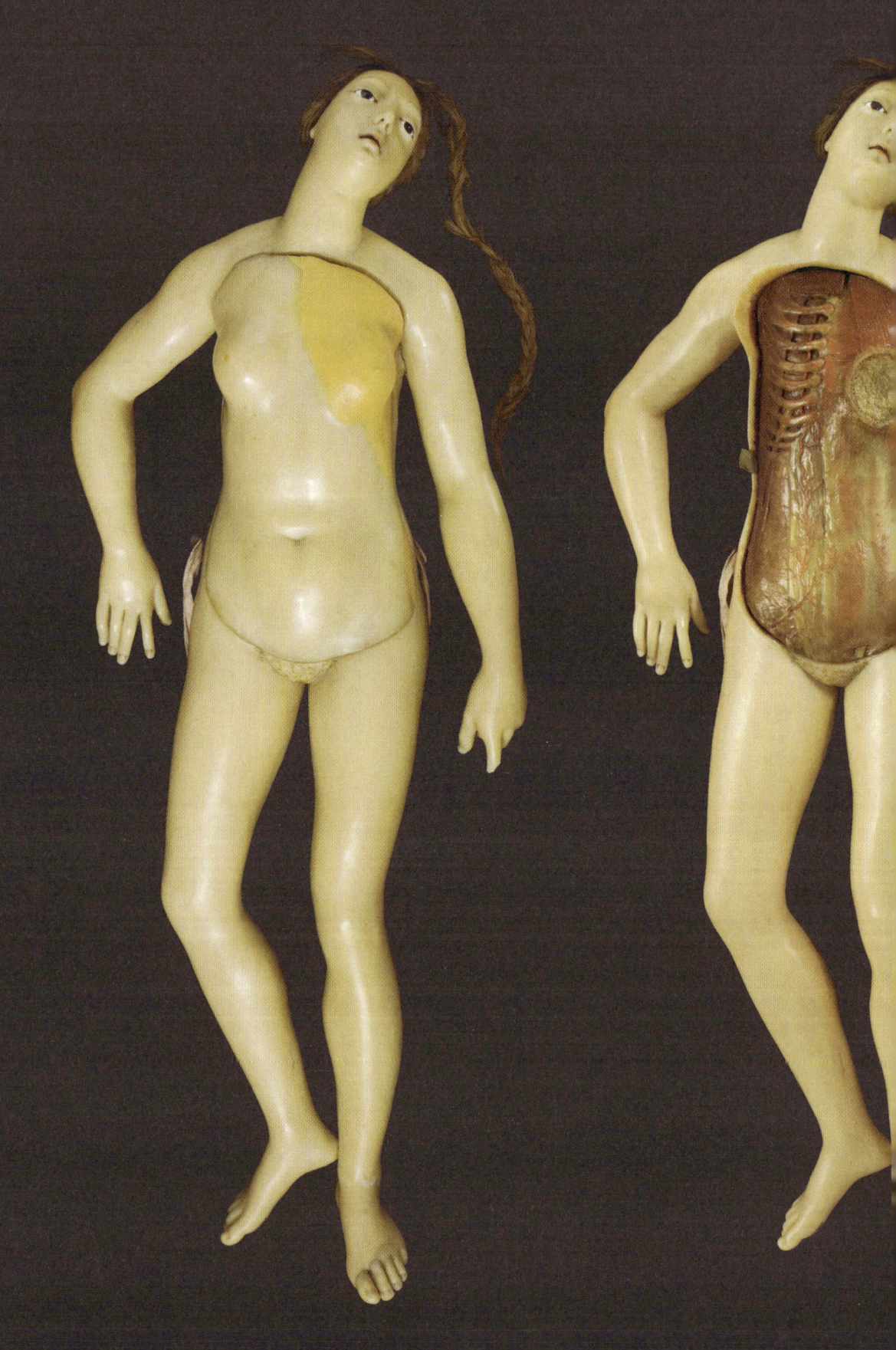

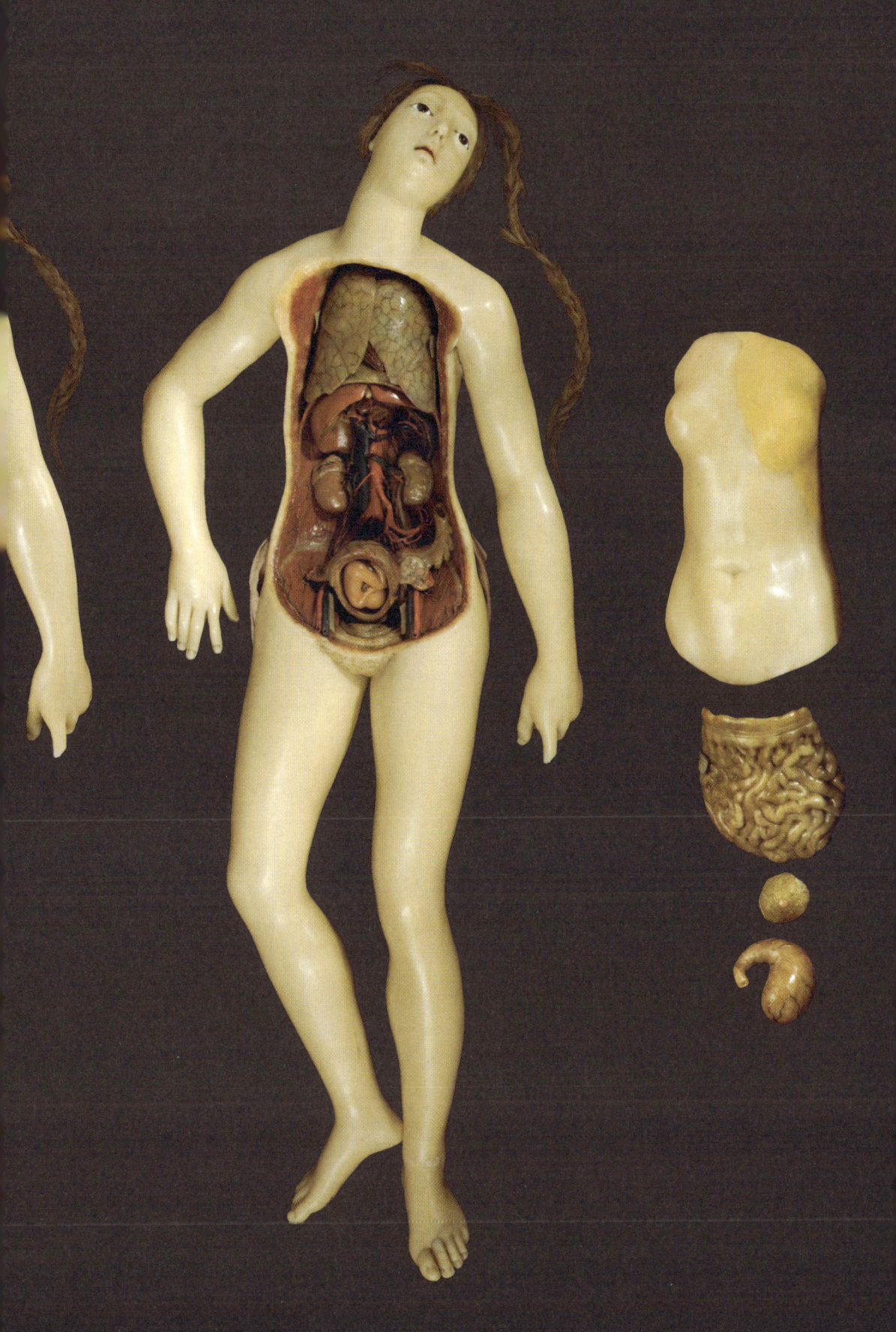

Wax model demonstrating the eye and optical nerves created by Clemente Susini between 1803 and 1805.

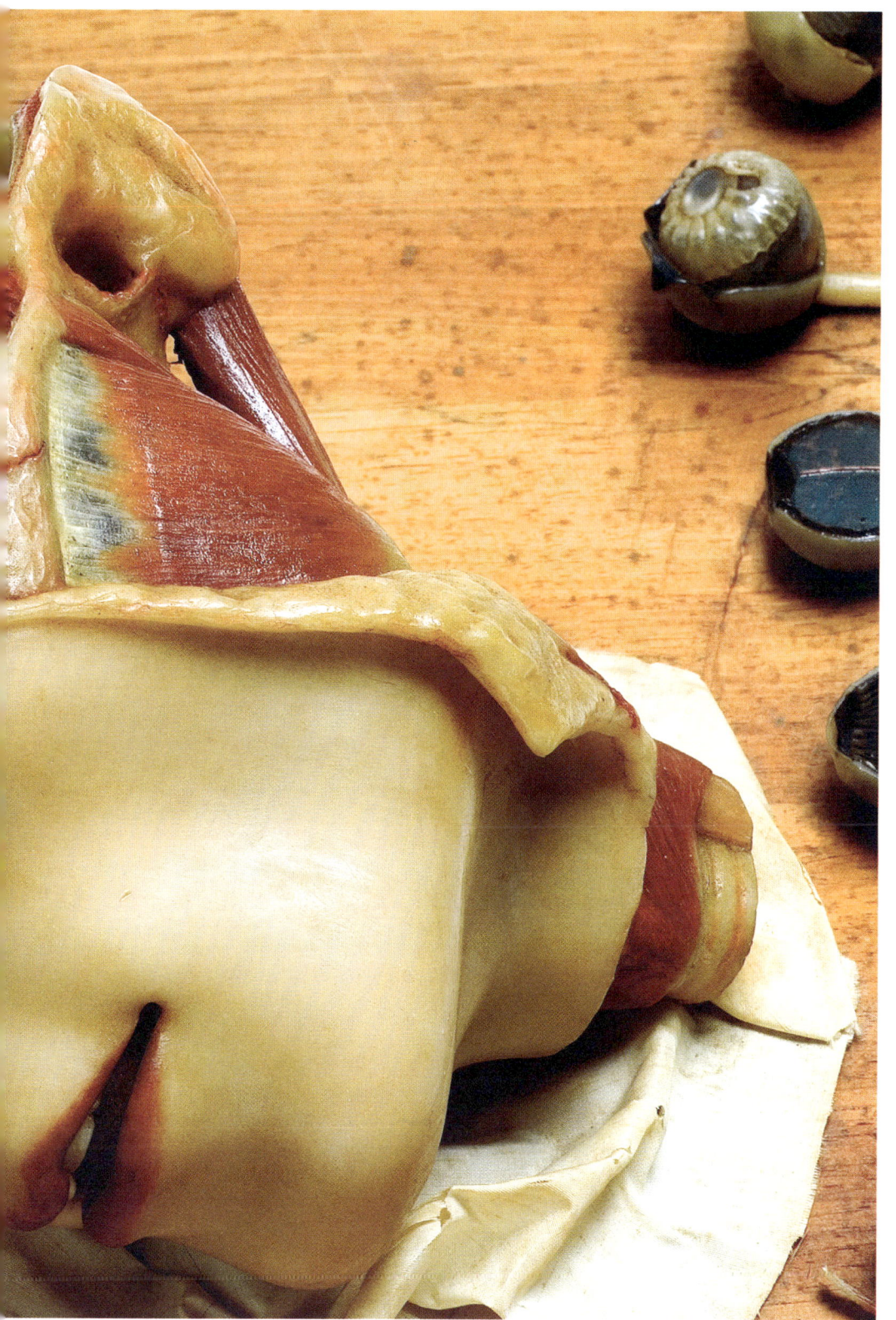

OVERLEAF A corpus sanctus—a manufactured body containing sacred relics—representing Saint Victoria, Santa Maria della Vittoria, Rome, Italy.

CHAPTER TWO

FROM·SACRED TO·SCIENTIFIC USE·OF·WAX

PREVIOUS

Effigy of Teresa Urrea (1873–1906), or Santa Teresita, Mexican folk healer and revolutionary. Her gift for miraculous healings was acquired after emerging from weeks spent in a trance. She lies in a golden casket at Carmen Alto, Oaxaca, Mexico, founded in 1696.

While Susini's expert artistry is responsible for the convincing appearance of La Specola's Medici Venus, she owes much of her unnerving presence to the wax with which she is made. Wax can look uncannily like flesh; it has a similarly moist appearance, depth of colour (due to the even suspension of added pigments), and transparent opacity. It has also been intimately related to death rituals, where it represents the stillness of a corpse that appears only to need its spirit to be immediately reanimated. Wax is, by nature, contradictory: solid and molten, stable and ephemeral, 'flesh' and yet simulacrum, seemingly alive, yet merely material.

Because of these qualities, wax has been the medium of choice for the making of human surrogates for anatomical, popular, religious, and magical

fig. 28

fig. 29

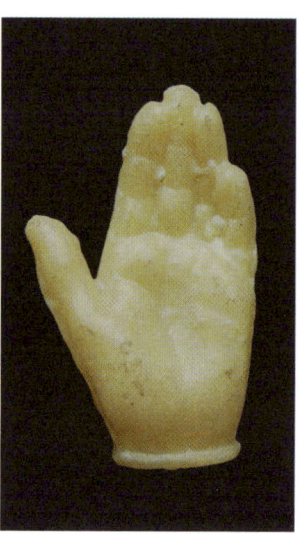

fig. 30

fig. 28 Small wooden coffin containing a beeswax poppet—with a slot in its back, into which written spells, nail parings, or hair, can be inserted—that once belonged to a clairvoyant known as Madam de la Cour.

fig. 29 This Agnus Dei—a wax amulet, imprinted with the lamb of God and blessed by the pope—was found in Oxfordshire, England, and dates from 1578, when it was a criminal offence to be Catholic.

purposes. Wax figurines in the Ancient Egyptian *Book of the Dead* (c. 1550–50 BCE) are described as engraved with the names of Ramesses III's enemies and bound with string. Voudou dolls and poppets, proxy bodies meant to inflict harm or death to enemies, have been used since at least the medieval period. Wax was also an essential part of the mummification process. The portraits that decorate the mummy casings at Fayum, Egypt, were painted in a combination of wax and pigment known as encaustic. They were displayed in the home during the life of the subject, and, upon death, attached to the mummy case in which their owner would spend eternity. In Ancient Rome, wax was used to create death masks and funerary effigies of notable personages, presaging the wax museums.

The symbolic and ritual significance of wax has been particularly rich in Christian, especially Catholic, tradition. Wax is seen as fragile, transitory, and malleable: like man, moulded by God and lit by his spirit. Bees are ascribed exemplary virtues in Christian iconography, and liturgical candles are traditionally made of beeswax. In Italy, the plague of 1575–77, which decimated the population, also caused the rapid expansion of the wax industry and greatly raised the price of wax, as the demand for candles for use at funerals was vastly increased. For many commoners beeswax was an expensive luxury only experienced at a funeral service; ordinary candles were made of tallow. The Roman Catholic tradition of Agnus Dei—wax amulets imprinted with an image of a lamb bearing a cross or flag, symbolizing Jesus Christ, and blessed by the pope— depends upon such associations, as do the anatomically themed wax votives known as ex-votos, or *boti*, which are left at a candlelit church or shrine to commemorate or request divine intervention. In the thirteenth century, St Francis of Assisi used a wax impression as a metaphor for the inner transformation of his religious experience, stating in the *Canticle of Love*: 'My heart softens like melted wax, and the form of Christ is traced upon it.'

fig. 30 Modern-day wax anatomical votive, made in Fatima, Portugal. Votives are used to request or offer thanks for healing. The shape reflects the ailing body part.

fig. 31

fig. 32

The physical body that is itself evoked by wax holds a special and seemingly paradoxical position in Christian belief. As the seat of the passions and of sin, the body must be tamed or mortified, yet in one of the central mysteries of the faith Christ was 'the Word made flesh': simultaneously man and God. Jesus lived and suffered in a human body that was given in the ultimate sacrifice, one that rendered all further animal sacrifice unnecessary. The sinful and fleeting pleasures of the body are believed to be mere vanities compared to the everlasting life that awaits the virtuous in heaven; yet the corporeal bodies of the faithful are essential to their salvation, and are believed to be resurrected and reunited with their souls in heaven when the trumpets sound on Judgement Day. Even in everyday religious observance, the body is central to the rite of the Eucharist, or Holy Communion, a commemoration of the Last Supper in which the believer eats a wafer and drinks wine that has been converted by the priest into the body and blood of Christ.

Predicated on such equivocal meanings of the body, the Catholic 'cult of the saints' concerns the desire to be close to the powerful physical remains of saints.

fig. 31 Francesco Stelluti's engraving from 1625 combines an early illustration of bees observed through a microscope with a Latin poem complimenting Pope Urban VIII, whose family's emblem was bees, and to whom the image was presented.

fig. 32 Beekeepers and the Birdnester (c. 1568) as depicted in pen and ink by Pieter Brueghel the Elder.

Egyptian death rituals are intimately related to wax: 'mummy' traces the Ancient Egyptian word 'moum', meaning 'wax' or 'tallow', which was an important part of the mummification process. Painted in encaustic, a combination of wax and pigment, the Fayum mummy portraits from Egypt are believed to date from between the late first century BCE and mid third century CE. About nine hundred portraits have been discovered at Fayum, near Cairo. Beautifully preserved, they are the oldest known body of art from classical antiquity. Each portrait would have been commissioned during the subject's lifetime and displayed in the home until their death, whereupon they would be attached to the mummy case.

LEFT   The mummy of an adolescent boy from Fayum, near Cairo, Egypt (c. 100 to 120 CE). The body was wrapped whilst in an advanced state of decomposition—the ribs and spinal column are in a state of confusion—with fine linen layers arranged in a diamond pattern with gilded studs. An encaustic (pigmented wax) mummy portrait has been inserted over the face.
ABOVE   This encaustic portrait on limewood from Fayum, Egypt, depicts a wealthy woman. Her jewelled necklace and diadem are reproduced in gold leaf.
OPPOSITE   Encaustic portraits from mummy cases found at Fayum, near Cairo, Egypt.

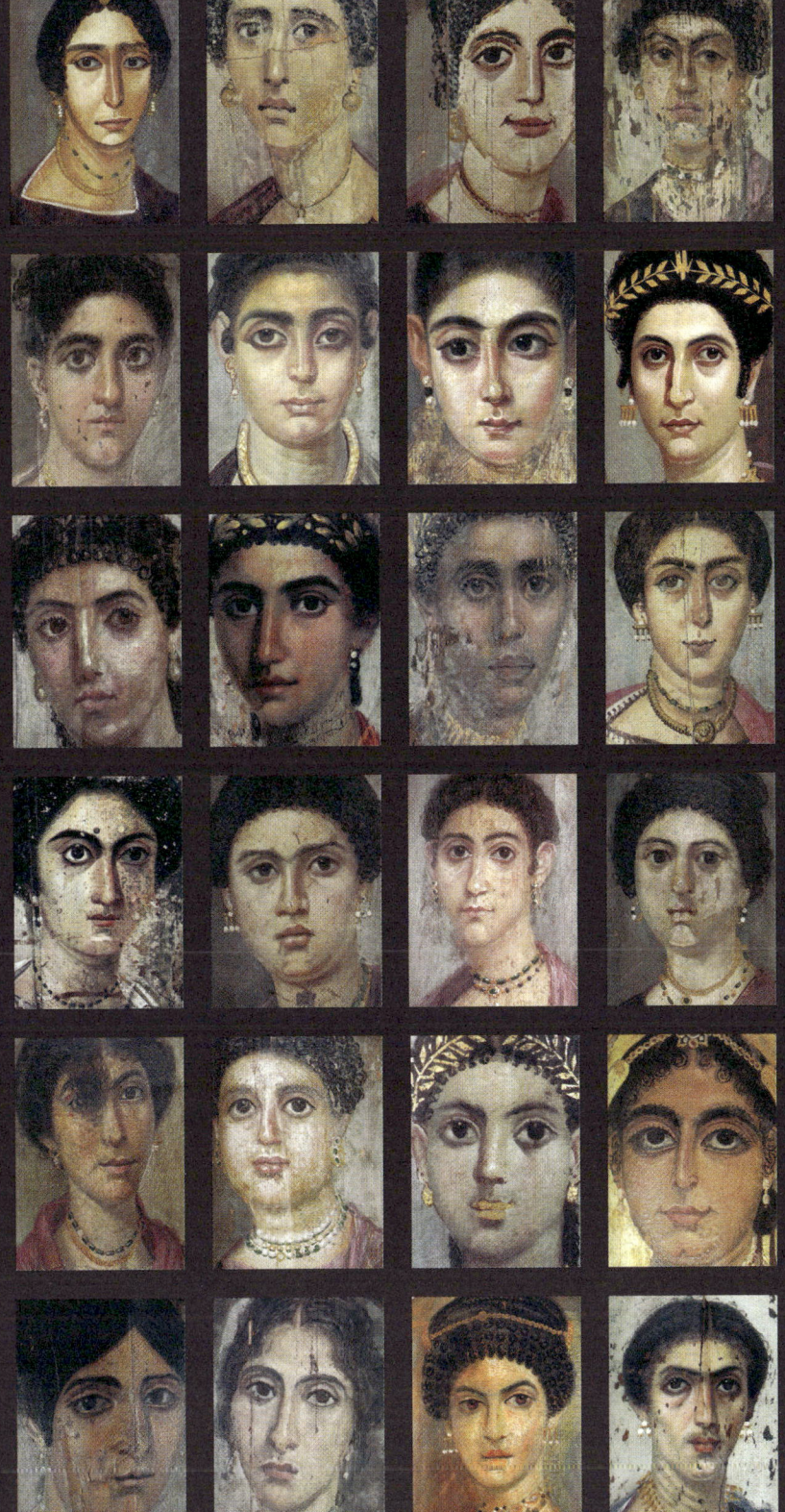

[2] Death mask of a young girl from Lugdunum, Roman Gaul—now Lyon, France. This plaster cast is the mould for a wax mask, which has not survived.

TO·THE
DEPARTED
SPIRIT·AND
IN·THE
MEMORY
OF·CLAUDIA
VICTORIA,
WHO·LIVED
TEN·YEARS,
ONE·MONTH,
AND·ELEVEN
DAYS.

This epitaph accompanying the death mask opposite further states: 'Her mother Claudia Severina made this monument for her sweet daughter and for herself in her lifetime.'

In Italy, before the rise of scientific medicine, many people's main strategy for coping with death and disease was miraculous healing. If you suffered from an ailment, there was a specific saint to whom you could pray or leave an offering: Saint Sebastian would be invoked against the plague, Saint Agatha for breast tumours, Saint Anthony for the pregnant, Saint Maurice for cramps, Saint Denis for headaches, Saint George for skin diseases and syphilis, and so on. According to medical historian Jacalyn Duffin, the vast majority of the miracles performed by saints were medical in nature.

Shrines were often built above the burial sites of martyrs (those persecuted, often enduring gruesome deaths, for their religious beliefs) so that much early Christian worship took place in catacombs; later, elaborate churches were built

fig. 33

fig. 33 Painted and gilded oak reliquary bust containing the skull of Saint Balbina, an early Roman virgin martyr (c. 1520–23).

on top of such tombs. Miraculous relics (usually a body part or personal effect of a saint) were, from the start, a powerful draw for pilgrims and tourists: in 787 CE an edict ruled that a Catholic church could not be consecrated without one. If no saint had been interred nearby, a fragment of the saintly body, an object directly related to the life of Christ, or even something touched by the saint or one of the relics associated with the saint, would suffice. These would be encased in purpose-made and often highly embellished vessels called reliquaries. The collection and display of relics and reliquaries became big business, and the relics themselves were a common spoil of war. Very often, the same relic can be found in several churches, suggesting that the urgency and importance of acquisition sometimes outweighed concern for authenticity.

Sometimes an entire preserved body might serve as a relic, as in the case of incorruptible saints, whose bodies, miraculously, do not decompose after death and are even sometimes reported to give off a sweet smell. Gaining status as 'incorruptible' varies in the extreme: Saint Catherine of Bologna was declared such after just eighteen days in the ground, while Saint Cecilia was buried for over seven hundred years before she was pronounced incorrupt. Many of the incorruptible saints seen in churches are mummified or skeletal, and some have been 'touched up' with wax masks and hands and are displayed in crystal caskets, much like clothed Anatomical Venuses.

In the wake of the devastation wrought by the Black Death in Europe after 1348—killing as many as two thirds of the population in some areas, and disproportionately affecting men between twenty-six and forty-five—Christian ritual began to focus more on suffering and death. This led to an increasing devotion to the body and blood of Christ in the ritual of the Eucharist and a cult of relics of the blood of Christ. The impact of the plague also renewed interest in the idea of purgatory, originally developed in the eleventh and twelfth centuries.

fig. 34 Ebony caskets, containing wax busts with gold and silver leaf, lapis lazuli, and enamel details, from Lombardy, Italy (c. 1600–1610). The busts portray, from left to right, a 'Soul Condemned to Purgatory', a 'Blessed Soul', and a 'Damned Soul'.

fig. 34

One of the most successful and contentious of Christian ideas, purgatory is held to be a state in which all humans, except martyrs, must spend time in the purgation of their sins before attaining the purity necessary to enter heaven. An industry of purgatory quickly grew, focused on prayers for those languishing there. Souls whose transit through purgatory had been expedited due to such prayers were expected to remember their benefactor once they reached heaven, and to intercede with God on their behalf. Even hospitals were part of the purgatory industry; those who could afford to aged and died at home, while hospitals were regarded as charity homes. Patients being cared for through the benevolence of others were expected to pray for the souls of their benefactors.

The tradition of purgatory lives on in popular worship in places such as Cimitero Delle Fontanelle in Naples, where a community of mostly older women take part in what is called 'The Neapolitan Cult of the Dead' or 'The Neapolitan Skull Cult'. They descend to the underground catacomb on Mondays, the day traditionally dedicated to Hecate, Greek goddess of the underworld. Monday is also the special day on which one of the many manifestations of the Madonna, Our Lady of Mount Carmel, is believed to be able to free souls from purgatory. The women traditionally choose an anonymous skull based on a dream, which believers understand as a mediation between the world of the here and the

THE·LIGHT
IS·GOD,·THE
WAX·IS·MAN,
CHRIST
IS·BOTH...
HE·IS·THE·WAX
THAT·MELTS
AND·ENDS·IN
DEATH,·LIKE
US·MORTALS.

Cornelio Musso (1511–74). Bishop of Bitonto. Quoted in Il quarto libro delle prediche del reverendissimo mons (1579). Styled the 'Italian Demosthenes', Musso was a highly distinguished orator.

Seventeenth-century, life-sized painted wooden effigy of the suffering Christ, displayed in the Chiesa dei Santi Filippo e Giacomo, Naples, Italy.

hereafter. They clean and care for their selected skull, and leave it offerings. Sometimes the skulls are decorated, or placed in customized shrines. Prayers are said for the abandoned souls—popularly called *le anime pezzentelle*—which are considered to possess powers approaching the saintly that enable them to intercede on the petitioner's behalf.

The purgatorial industry reached its zenith in the sixteenth century, when the proceeds from selling indulgences—papal grants promising to shorten or cancel a person's time in purgatory, which in the twelfth and thirteenth centuries were sold as ubiquitously as lottery tickets—were blatantly used to fund the rebuilding of St Peter's Basilica in Rome. On 31 October 1517, the German Augustinian friar Martin Luther (1483–1546) is said to have nailed his 'Disputation...on the Power and Efficacy of Indulgences'—also known as the 'Ninety-Five Theses'—to the door of All Saints' Church in Wittenberg. In his

fig. 35 Early-twentieth-century photograph of life-sized wax ex-voto effigies in the Basilika Vierzehnheiligen, Germany.

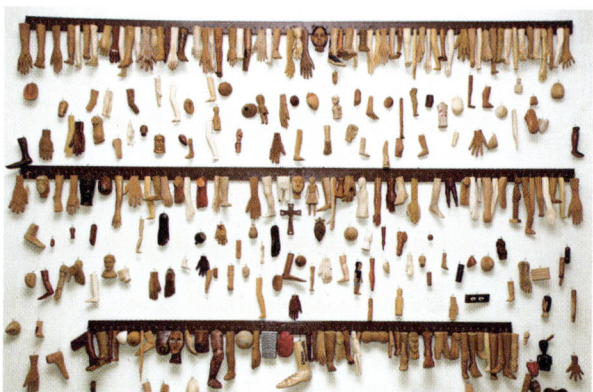

fig. 36 Anatomical ex-votos, or milagros ('miracles'), as they are known in Latin America, left as offerings at the Casa dos Milagres, or House of Miracles, a shrine to Saint Francis in Canindé, Brazil.

manifesto, Luther condemned the selling of indulgences to raise church funds from the pockets of the poor, and criticized the Catholic cult of the saints for having no foundation in the Bible and amounting to the worshipping of false gods. Luther's demands for reform led to a schism in the Christian Church, with his followers establishing new Protestant churches, free of the pope's rule and with an unmediated relationship to God via the literal word of the Bible.

Catholics responded to the Protestants' denunciations with a series of reforms made between 1545 and 1563. These ushered in what is termed the Counter-Reformation, in which the practices that were most heavily criticized by the Protestants—the cult of the saints, the magical use of relics, and the veneration of the Virgin Mary—were strongly reaffirmed, although perceived abuses such as the wanton selling of indulgences were abolished. This reaffirmation of the importance and value of saints, and of a sacred material culture, had profound implications for Catholic rites and art.

As the cult of the saints became increasingly important to Catholic worship in the wake of the Counter-Reformation, relics and sacred representations of the saints proliferated. Churches commissioned a variety of life-sized and extremely life-like effigies representing saints and martyrs made of painted wood, plaster, or tinted wax. Effigies might be combined with relics, such as in the wax representation of blessed Imelda Lambertini, patron saint of 'fervent first communions', still on display at the Church of San Sigismondo in Bologna; her bones are stored behind gilt-framed glass panels beneath her body. They might also be combined in the form of a 'corpus sanctus', in which a body manufactured of wax or other media also acts as a reliquary to hold the mortal remains of the saint. One such example is Saint Victoria, on view at the church of Santa Maria della Vittoria in Rome (see page 66). To this day, many churches in Italy, Portugal, Spain, and the former Spanish or Portuguese colonies still feature life-like manufactured saints. These religious effigies had a great impact on the visual language of the scientific anatomical wax models that followed.

One bizarre flowering of the Catholic fetishization of the corporeal body following the Counter-Reformation is found in what are known as 'sacred

fig. 37

fig. 38

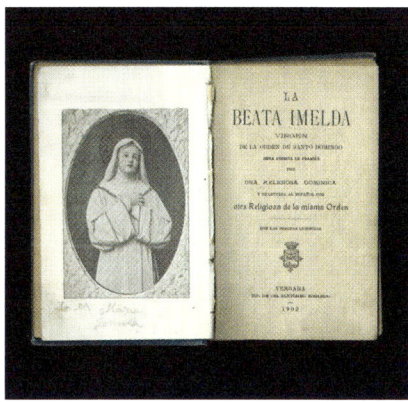
fig. 39

representations'. These were spectacular tableaux staged by 'confraternities of the dead', religious groups devoted to providing proper burials and death rites for the poor and abandoned. These charitable brotherhoods took care of the burial and prayers for the souls of the deceased poor in order to speed their time through purgatory. The Church of Santa Maria dell'Orazione e Morte in via Giulia, Rome, produced sacred representations from the eighteenth century until at least 1880; they took place in a cemetery beneath the church, which had been artistically decorated with skulls and bones. Life-sized wax figures were posed in tableaux dramatizing scenes such as 'The Death of Abel', 'The Martyrdom of St Arasmus' (complete with split belly and spilling entrails) and 'The Mountain of Purgatory', in which actual human corpses were used to represent souls being rescued from the flames of purgatory. Nearby, at Santo Spirito in Sassia, an 1813 tableau depicting the Last Judgement featured a wax angel playing a trumpet to raise the dead, represented by the fresh corpses drawn from the adjacent cemetery.

The power of sacred representations relied upon the realistic and literal nature of their depictions. Between 1200 and 1600, Florence was renowned for the quality of its wax anatomical votives, many of which were manufactured

figs 37 & 38 Twentieth-century devotional prayer cards, or 'santini', honouring Blessed Saint Imelda, who is also referred to as 'La Beata Imelda Lambertini'. As an adolescent, Imelda fell into a fatal swoon of overwhelming joy at her first communion with Jesus Christ. She is now the patron saint of first communions.

fig. 39 A prayer book based on devotions to, and autobiographical details of, Blessed Saint Imelda.

EXTENDING UPWARDS AND FAR ALONG THE WALLS AND CEILING THERE WAS A MULTITUDE OF BEINGS WHICH JOSTLED ONE WITH THE OTHER TO BE NEAR THE POWERFUL MADONNA. THIS WAS THE THRONG OF THE VOTIVE WAX IMAGES, THE EFFIGIES OF GREAT PERSONS IN FULL DRESS…

Passage describing the ex-votos at Florence's Santissima Annunziata. Taken from George Eliot's novel Romola (1862), in which the young Romola encounters religious, intellectual, and artistic culture in Florence.

THRUST·IN·THEIR
MIDST·THERE·WERE
DETACHED·ARMS
AND·OTHER·LIMBS...
IT·WAS·A·VERITABLE
MULTITUDE·OF
MORTAL·REMAINS
AND·FRAGMENTS
WHICH,·IN·THEIR
VARYING·DEGREES
OF·SQUALOR
RELIEVED·HERE·AND
THERE·BY·A·DASH
OF·COLOUR,
REFLECTED·THE
UNDERLYING·CROWD.

OPPOSITE *Illustration from an 1894 edition of the French daily newspaper* Le Petite Moniteur Universel. *It shows visitors to the Capuchin crypt in Rome on 1 November, All Saint's Day, which is dedicated to all the saints of the church—known and unknown.*

ABOVE *Photographs of nineteenth-century 'sacred representations', including (top left) a tableau dramatizing the vision of Ezekiel (1868), (centre left) a scene depicting the martyrdom of Diodorus and Mariano (1865), and catacombs decorated with bones and skulls.*

Offerings left at the Cimitero Delle Fontanelle, an underground catacomb in Naples, Italy, which serves as the focal po'nt of 'The Neapolitan Cult of the Dead' or 'The Neapolitan Skull Cult'.

for pilgrims paying homage to a miraculous painting of the Madonna at the Basilica della Santissima Annunziata. These votives would take the shape of an afflicted body part, as detailed by art historian Roberta Ballestriero. The visitors would leave their symbolic offerings at the shrine of the Madonna or other saints, in order either to request or to commemorate a miraculous intervention. Wealthier visitors would commonly commission a life-sized effigy of themselves, often dressed in their own clothes, leading Santissima Annunziata to function as a sort of sacred Madame Tussauds, where the devout and the curious alike could see realistic wax representations of famous and important individuals.

As the space became increasingly overstuffed and wax figures began to fall onto the heads of the devout from their frayed and rotting ropes, the effigies

fig. 40

fig. 41

fig. 40 Life and Death: a memento-mori oil painting of a young Italian woman from the seventeenth century.

fig. 41 Vanitas oil painting, depicting a woman half in life, half as a skeletal angel of death bearing a scythe.

fig. 42 An eighteenth-century, life-sized wax head, one side resembling Queen Elizabeth I, the other a skull teeming with insects and reptiles.

were phased out. The last of them were melted down to make candles in 1786 after Leopold II called for their destruction in 1785. As he explained, 'the votives obstruct the beautiful pictures there, and are a nest for dust'. Leopold II also curtailed popular religious processions—which often featured human remains in the form of relics—and other religious and popular entertainments and spectacles, which he saw as encouraging 'useless dissipation'.

The close, mimetic observation of the natural world combined with a metaphysical intention is also found in memento mori—artworks intended to remind viewers that they will die, in order to encourage them to lead a virtuous life and so to meet their maker after death. Memento mori were particularly popular in the seventeenth century, probably in response to the plague that continued to devastate Europe, and which still bore intimations of divine wrath. In Naples, where the Great Plague of 1657 wiped out half the city's population, the explicit wax tableau of *La Donna Scandalosa* (The Scandalous Woman) can be seen, equal parts natural study and moral exhortation.

Dutch artist, anatomist, physician, and preparatory and master embalmer Frederik Ruysch (1638–1731) exploited the religious and proto-scientific overlap. Famed for his mastery of wax injections in the creation of life-like specimens, Ruysch is best remembered today for his tableaux, which were equal parts memento mori and didactic objects. These extraordinary pieces were crafted from real human fetal skeletons (to which his role as master midwife gained him access), gallstones, and bits of human tissue. They were accompanied by allegorical texts, along with symbols such as mayflies, pearls, candles, and wreaths of flowers, in reference to the brevity of life. Rosamond Purcell and Stephen Jay Gould discussed Ruysch's work in their book *Finders, Keepers: Eight Collectors*, published in 1992. They recorded that one fetal skeleton holding a

fig. 42

fig. 43

fig. 43 La Donna Scandalosa, a late-seventeenth-century memento-mori wax and cloth piece from Oratorio Compagnia dei Bianchi Della Giustizia, Naples. It was created to serve as a warning to women who have led a dissolute life.

string of pearls in its hand proclaims, 'Why should I long for the things of this world?' Another, playing a violin with a bow made of a dried artery, sings, 'Ah fate, ah bitter fate.' Ruysch exhibited these in ten cabinets in five rooms at his home, which he opened to visitors twice a week.

Ruysch also displayed a 'reconstructed tomb' assembled from numerous bones and skulls, with a centrepiece featuring an embalmed fetus crowned with a floral wreath, surrounded by ten skeletons of adults, children, and fetuses, each bearing emblematic attributes—such as a trumpet, lance, or toy—or memento-mori-themed banners. The allegorical and artistic sensibility did not eclipse Ruysch's scientific aim: each object, down to individual gallstones, was carefully described.

Ruysch sold his collection to Peter the Great (1672–1725) in 1717 and much of it, including his famed tableaux, is now sadly lost. The closest we will probably ever get to what these tableaux might have looked like, aside from Cornelius Huijberts's contemporary engravings, is a 'Macabre Altar' from around the

Public dissection at the anatomical theatre in Leiden (c. 1609). Note the skeletons brandishing banners with phrases such as 'Pulvis et Umbra Sumus' ('We are [but] dust and shadow'), and the skeletal Adam and Eve.

fig. 44 Homo ex humo *(Man from the ground, or dust)*. This meditation on the creation of man incorporates both biblical and scientific perspectives. The creation of Adam by God is depicted in the centre, framed by illustrations of fetal development, almost certainly based on Ruysch's tableaux (see page 95).

same time. Fashioned from a mummified child and three fetal skeletons, and augmented by verses from Virgil and Malherbe, it was made by an unknown artist in the eighteenth century. Described in an early guide to the cabinet of Jean-Joseph Sue (its owner) as 'a sort of allegoric tomb displayed in a glass showcase', this enigmatic object is thought to be a devotional item: either a traditional memento mori or a 'profane relic'.

Ruysch studied—and perhaps found inspiration for his constructions—at Leiden's anatomical theatre in the Netherlands, which first opened in 1594. There, real human skeletons held aloft banners emblazoned with Latin proverbs such as *'Vita humana lusus'* ('Man's life is but a game') and *'Volat irrevocabile tempus'* ('Time flies and cannot be recalled'). At the centre of the theatre, a skeletal Adam and Eve reminded the viewer that their original sin in the Garden of Eden introduced the hateful concept of death to humanity.

fig. 44

fig. 45

A similar intermingling of memento mori and science, art and medicine is apparent in the work of masterful wax artist Gaetano Giulio Zummo (1656–1701), better known as Zumbo. Born to a noble family in Syracuse, Sicily, Zumbo studied with Jesuits and, despite never practising as a priest, ultimately took the title of abbot. He apprenticed in wax sculpture in Sicily, where ex-voto offerings and miniature nativity or 'crib' scenes in wax are still produced in a tradition stretching back to the late Middle Ages. Like the great artists of the Renaissance, it is thought that he studied anatomy, in the words of his early anonymous biographer, 'in order to imitate nature more accurately in her own perfect forms'.

Zumbo was renowned for a series of wax miniature tableaux, or *teatrini*. Now designated his 'Theatres of Death', they bear titles such as *Il Trionfo del Tempo* (The Triumph of Time), *La Peste* (The Plague), *La Vanità della Gloria Umana* (The Vanity of Human Glory), and *Il Morbo Gallico* (The French Disease—that is, syphilis). These pieces are also collectively referred to as his plague waxes, and although only one of them takes the plague as a direct subject, they nonetheless eloquently express the havoc wrought by the bubonic plague throughout Italy, especially in sixteenth-century Milan and Naples. These small-scale dioramas are filled with miniature human bodies exactingly rendered in tortured poses or various stages of decomposition, set into ruined landscapes. In one, 'Time'—personified as a winged old man bearing a scythe with a miniature framed portrait of the artist at his feet—is the only live figure, surrounded by putrefying babies, skeletal parts, and a swooning, bare-breasted woman. *La Vanità della Gloria Umana*

fig. 45 *Engraving by Cornelius Huijberts from Frederik Ruysch's* Thesaurus Anatomicus *(1703). The tableau shown was crafted from fetal skeletons, gallstones, and preserved human tissue. The reclining skeleton holds a mayfly, symbolic of the brevity of life.*

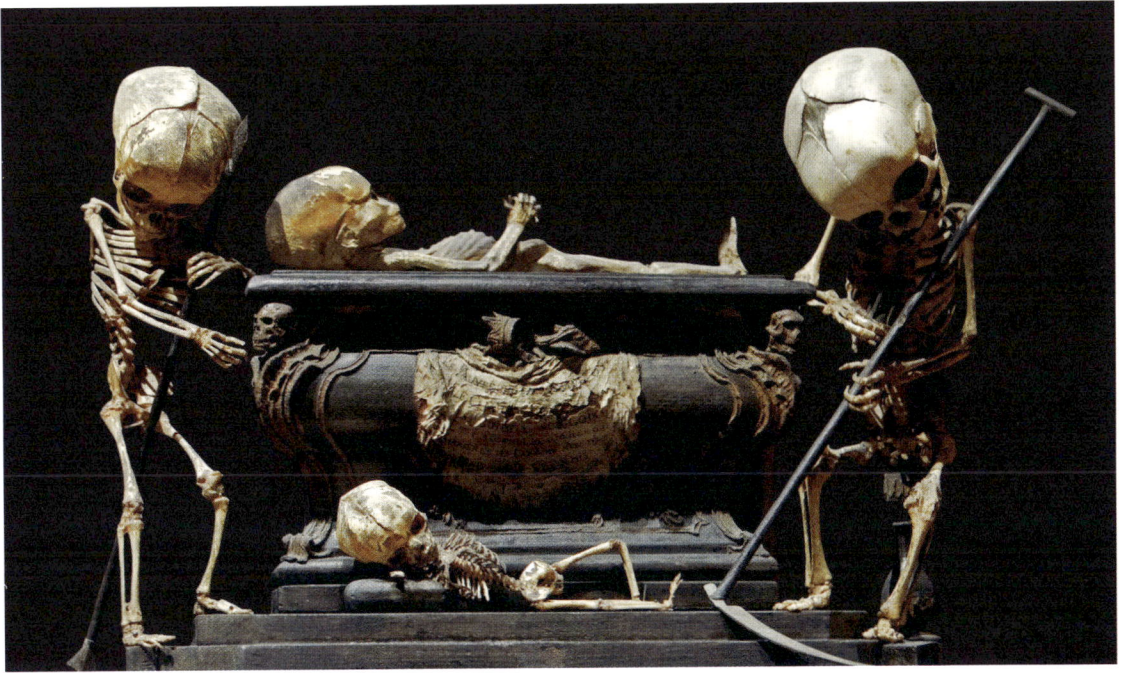

fig. 46

depicts a decaying cemetery, where a tomb sculpture of a pensive woman looms above bloated and rotting bodies in poisonous tones of green and brown, fed on by rats and insects. The surviving figures of *Il Morbo Gallico*—of which, sadly, only fragments remain after the great Florentine flood of 1966—include a man howling in agony and a memorable metaphor of the fatal, sexually transmitted disease: the blindfolded figure of a cupid with his belly cut open and his tiny entrails pouring out.

Zumbo's dramatically realistic material drew varied interpretation. The Marquis de Sade (1740–1814) was an ardent admirer of Zumbo's Theatres of

fig. 46 *Late-seventeenth-century 'Macabre Altar', or 'profane relic' fashioned from a mummified child and three fetal skeletons, from the cabinet of Jean-Joseph Sue, French surgeon and anatomist.*

Illustration by Cornelius Huijberts of a tableau by Dutch anatomist and artist Frederik Ruysch, from a guidebook to his home museum. Ruysch constructed a series of such tableaux—using fetal human skeletons, gallstones, and hardened veins and arteries—which were both scientific objects and memento mori. The skeleton on the right weeps into a 'handkerchief' crafted from injected mesentery or brain meninges. Ruysch's tableaux were ultimately purchased by Peter the Great for his collection in Russia, but are now sadly lost.

ad vivum Sculpsit

Death. De Sade saw them while in Florence, not on a Grand Tour but on the run from French charges of sodomy and poisoning. To him, they reflected his own preoccupations with sex, cruelty, and death (see pages 104–105). But the American author Herman Melville (1819–91) wrote of 'horrible humiliation' in his *Journal Up the Straits* (1856–57) and proclaimed Zumbo a 'Moralist, this Sicilian'.

Zumbo's Theatres of Death combine a didactic theological intent with a highly naturalistic, observation-based sense of scientific inquiry, creating a tension that is one source of their special power. This tension is related to a specific tradition in Europe at the end of the seventeenth century, when meditation on the decomposition of the body in the tomb with as much detail as possible was recommended as a spiritual exercise. This was suggested in Jesuit Daniello Bartoli's (1608–85) popular devotional manual *L'huomo al punto, cioe l'huomo al punto di morte* (Man at the Turning Point, that is, Man at the Point of Death) (1667), which included a chapter entitled 'The tomb a school able to make even the mad wise: we enter therein to hear a lesson of moral and Christian philosophy'.

fig. 47

fig. 47 Il Morbo Gallico (The French Disease), a Theatre of Death by Zumbo. His patrons included Cosimo Medici and his son, Ferdinando. Ferdinando contracted syphilis, probably at Venice's Carnival of 1696, which perhaps inspired this piece. Note the blindfolded figure of a decomposing cupid: an apt, if graphic, symbol of the disease itself.

Zumbo's work drew much acclaim, and came to the attention of Grand Duke Cosimo III (1642–1723), an extremely pious man and the penultimate Medici to rule Tuscany. Delighted by his skill, Cosimo became Zumbo's patron and granted him a generous pension. Zumbo also produced work for Cosimo's son, the Grand Prince Ferdinando (1663–1713), who, unlike his father, was a great lover of arts, music, and liaisons. He caught syphilis, probably at the Venice Carnival of 1696, which perhaps inspired Zumbo's tableau devoted to the disease.

Around 1700, Zumbo was approached by French surgeon and anatomy professor Guillaume Desnoues. Desnoues wished to hire Zumbo to create a life-sized wax likeness of an important medical preparation that was beginning to decompose: a woman who had died during labour along with her unborn fetus. This commission, which Desnoues intended to secure him a French royal warrant, instigated a short-lived partnership between the two men. They were to part ways soon after completion of the anatomical model, after a quarrel that, some conjecture, concerned ownership of the invention of the technique of anatomical wax modelling. Desnoues is also said to have complained that Zumbo had created a 'half-putrefied corpse in wax' which could only 'awaken in the mind of the spectator...all the horrors of the tomb', while Desnoues's goal had been to commission 'the anatomy of the human body in relief without exciting the feeling of horror men usually have upon seeing corpses'.

Zumbo went on to create a few more anatomical waxworks, including two beautiful heads, one of which can still be seen today at La Specola, before leaving Italy for Paris. There, Louis XIV (1638–1715) awarded him the warrant to be the sole manufacturer of anatomical ceroplastics that Desnoues had so coveted, but, sadly, he died only a year later. Desnoues found a new partner in François de la Croix (1653–1713), an ivory carver with whom he collaborated in the early eighteenth century on 'Anatomies in waxwork'. After displaying the collection in Paris, the pair took the anatomical waxworks on tour around cities in England and France. Among other exhibits, the collection included '...the anatomy of a woman to the waste [sic] where all parts of the brain may be seen, and taken out of their place, and set back again.' (Haviland and Parish, 1970).

fig. 48 Zumbo's La Vanità della Gloria Umana *(The Vanity of Human Glory).*

fig. 48   *fig. 49*   *fig. 50*

Beyond simply creating the first anatomical model, the collaboration between Zumbo and Desnoues set the language for wax anatomies to come in the creation of idealized wax bodies that seemed to be either freshly dead, in a state of suspended animation, or alive. This approach was pivotal in making anatomical study more acceptable and appealing, and less frightening or disgusting, to the general public. It gave rise to a flowering of anatomical models in Italy, mostly created for public edification and entertainment, up until the early twentieth century.

As the leaders of Italy's church and state—Leopold II and Pope Benedict XIV—wished to discourage what they deemed the more superstitious practices of Catholicism, they hired the Church's most skilled ceroplasticians to create a new, rational society. The pope, born Prospero Lorenzo Lambertini and also known as the 'Enlightenment Pope', reigned from 1740 until his death. He was popular and beloved, known for his sense of fairness and for being kind, witty, and immensely learned. He supported the arts, the rights of women within the academy, and even corresponded regularly with Voltaire, who dedicated his play *Mahomet: A Tragedy* (1736) to him. Moreover, Pope Benedict XIV avidly promoted experimental science, medicine, and the study of anatomy. Before the middle of the eighteenth century, he had relaunched the existing Institute of Science at Bologna's Palazzo Poggi to create the world's first anatomy museum.

fig. 49 Zumbo's Il Trionfo del Tempo *(The Triumph of Time). The only living figure is that of Time, featured as a winged old man bearing a scythe.*

fig. 50 Zumbo's La Peste *(The Plague). His tableaux are sometimes referred to as plague waxes, although this is the only one that overtly deals with the havoc wrought by the bubonic plague in sixteenth-century Italy.*

[2] Zumbo's La Vanità della Gloria Umana (The Vanity of Human Glory) depicts a decaying cemetery in which a tomb sculpture of a pensive woman looms above bloated and rotting bodies fed on by rats and insects.

WHAT·GREATER
SUCCESS·CAN
ONE·WISH·FOR
IN·THE·ARTS
AND·SCIENCES
THAN·TO
FIND·THE
SECRET·OF
IMITATING·THE
WORKS·OF·THE
CREATOR·BY

Excerpt from the letters of French professor of anatomy and surgery Guillaume Desnoues (1650–1735), which were first published in 1706.

DEMONSTRATING
THE·ANATOMY
OF·THE·HUMAN
BODY·IN·RELIEF
WITHOUT
EXCITING·THE
FEELING·OF
HORROR·MEN
USUALLY·HAVE
ON·SEEING
CORPSES?

Detail of Zumbo's Il Trionfo del Tempo (The Triumph of Time). A miniature painted portrait of the artist rests at the heel of the figure of Time.

SO·POWERFUL·IS
THE·IMPRESSION
PRODUCED·BY
THIS·MASTERPIECE
THAT·EVEN·AS·YOU
GAZE·AT·IT·YOUR
OTHER·SENSES
ARE·PLAYED·UPON;
MOANS·AUDIBLE,
YOU·WRINKLE
YOUR·NOSE·AS
IF·YOU·COULD
DETECT·THE

Donatien Alphonse François, better known as the Marquis de Sade, describes his reaction to Zumbo's realistic 'Theatres of Death' in his novel Juliette, or Vice Amply Rewarded (1797).

EVIL ODOURS OF MORTALITY... THESE SCENES OF THE PLAGUE APPEALED TO MY CRUEL IMAGINATION: AND I MUSED, HOW MANY PERSONS HAD UNDERGONE THESE AWFUL METAMORPHOSES THANKS TO MY WICKEDNESS?

One of the main legacies of Pope Benedict XIV's tenure—and a major aim of his museum—was the encouragement of a more 'rational' Catholicism. He advocated the study of science as a way to pay homage to the divine and, radically, implored his clergy to encourage their flocks to donate the bodies of their kin, 'dead by whatever means', for dissection. His *De Servorum Dei Beatificatione et Beatorum Canonizatione* (On the Beatification and Canonization of the Saints) outlined the ways in which scientific methodology could be used to determine whether an alleged miracle was true or false, and laid down new guidelines for the beatification and canonization of saints. He held that religion and science coexisted as the two means of arbitrating human relationships with death, health, and disease.

The Institute of Science at Palazzo Poggi had been founded in 1711 by General Luigi Marsili (1658–1730). It represented the sum of scientific knowledge at the time, and aimed to encourage experimental research and establish new connections between previously separate fields of study. Artistic imagery was used to harmonize and unify the data that the experimental method had dissected

fig. 51

and interpreted across many exhibits, disciplines, and practices. Pope Benedict XIV developed and expanded the Institute, and in particular provided the Anatomy Room, along with its anatomical wax models whose purpose was to teach artists, medics, and the general public about the divinely rational structure of the human body.

In 1742, Pope Benedict XIV commissioned artist Ercole Lelli (1702–66) to create the waxworks that would provide the highlight of his museum. Lelli had already made his name, not only as a traditional painter of religious themes, but also as an anatomical artist, with two life-like wax kidneys and two wooden écorché figures he made for Bologna's anatomical theatre. Like many artists before him, he was known to have undertaken a number of his own dissections in the service of his craft. Lelli was helped by surgeon Giovanni Manzolini (1700–55). Giovanni's wife Anna Morandi Manzolini (1716–74) eventually surpassed her husband in skill and renown, becoming a professor of anatomy and visiting the court of Catherine II of Russia. The display, which can still be seen today, showcases Lelli's series of life-sized standing wax figures of both genders, built on human skeletons and demonstrating various levels of muscular dissection. These are flanked by his pair of fully embodied, naked waxworks, with real human hair and glass eyes, representing Adam and Eve. The final piece on each side of the room is a bare human skeleton holding a scythe: an angel of death.

Lelli's display promoted a scientifically accurate understanding of human anatomy based on dissection and, at the same time, encouraged a meditation on death as mankind's legacy, stemming from Adam and Eve's original sin in the Garden of Eden. In the tradition of memento mori, the skeletal angels of death terminating each progression remind one of the brevity of life. Equal parts sacred representation and instructional display, the exhibition challenges modern lines of demarcation between church and collection, memento mori

fig. 51 Panoramic view of the Anatomy Room at the Museum of Palazzo Poggi, Bologna, Italy, commissioned in 1742 by Pope Benedict XIV.

and science lesson, body and soul, relic and specimen. It also provided direct inspiration for the wax workshop at La Specola to create the Medici Venus.

As religious leaders began to discourage the practices of the cult of the saints, relics, ex-votos, and Agnus Dei, the need for wax modellers decreased substantially. Many of the finest artists of the votive trade were instead hired to craft the first anatomical wax models. Clemente Susini himself had made at least two wax effigies of the dead Christ for churches before being hired to fashion models for Fontana at La Specola. Cesare Bettini (1801–55), the creator of the wax effigy of Blessed Saint Imelda, went on to craft wax figures for the University of Bologna. It is little wonder that the wax figures these artists created seem to have existed precisely at the point where the sacred and the scientific converged.

A comment on the increasingly naturalistic memento mori of the Baroque era, from Michael Snodin and Nigel Llewellyn (eds), Baroque: Style in the Age of Magnificence 1620–1800 *(2009).*

[2]

A·NEW,·SCIENTIFIC
INTEREST·IN·THE·BODY
IS·DISCERNIBLE...·THIS
REALISM·IN·DEPICTING
THE·HUMAN·ANATOMY
INTRODUCES·A·TENSION
THAT·COULD·BE
INTERPRETED·AS·PART
OF·A·SHIFT·IN·THE
EXPLANATION·OF·THE
NATURE·OF·THE·BODY
FROM·THEOLOGY·TO
NATURAL·SCIENCE...
A·VIRTUOSITY·IN
SIMULATING·THE·'TRUTH'
OF·THE·MATERIAL
WORLD·HAD·THE
PARADOXICAL·FUNCTION
OF·MAKING·SUCH
IMAGES·MORE·EFFECTIVE
TRANSMITTERS·OF
SUPERNATURAL·POWER.

[2] The life-sized wax figures of Adam and Eve by Ercole Lelli that were commissioned in 1742 by Pope Benedict XIV for the anatomical wax museum in Bologna, Italy. (109)

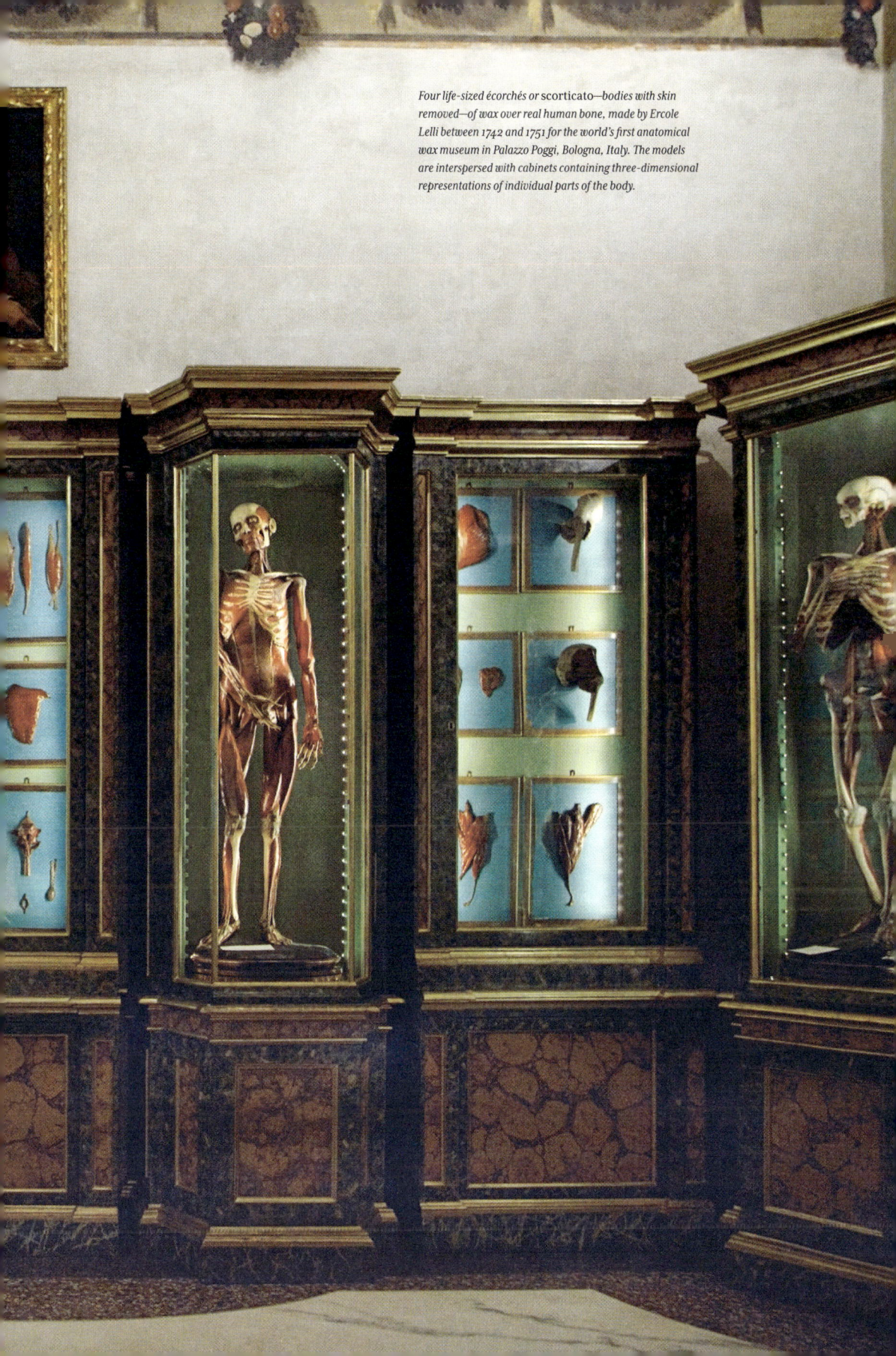

Four life-sized écorchés or scorticato—*bodies with skin removed—of wax over real human bone, made by Ercole Lelli between 1742 and 1751 for the world's first anatomical wax museum in Palazzo Poggi, Bologna, Italy. The models are interspersed with cabinets containing three-dimensional representations of individual parts of the body.*

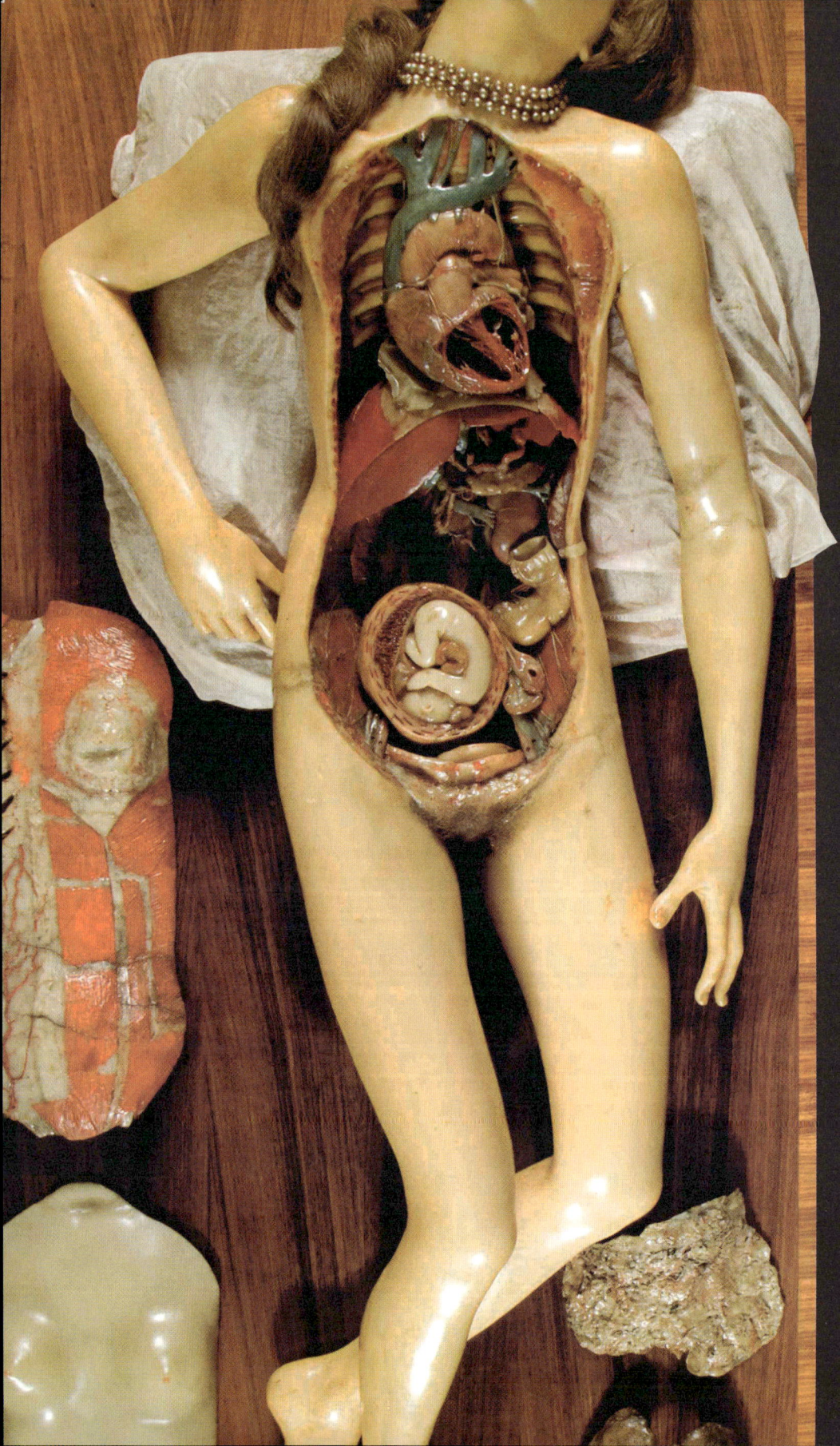

The Venerina (Little Venus) is a dissectible wax Anatomical Venus, made by Clemente Susini for the Museum of Palazzo Poggi, Bologna, Italy (1782). Although partly modelled on La Specola's Medici Venus, the Venerina is smaller in scale. The museum, which still displays the model, describes it as follows:

'The agony of a young woman is represented in her last instant of life as she abandons herself to death voluptuously and completely naked.' Her 'thorax and abdomen can be opened, allowing the various parts to be disassembled so as to simulate the act of anatomic dissection.'

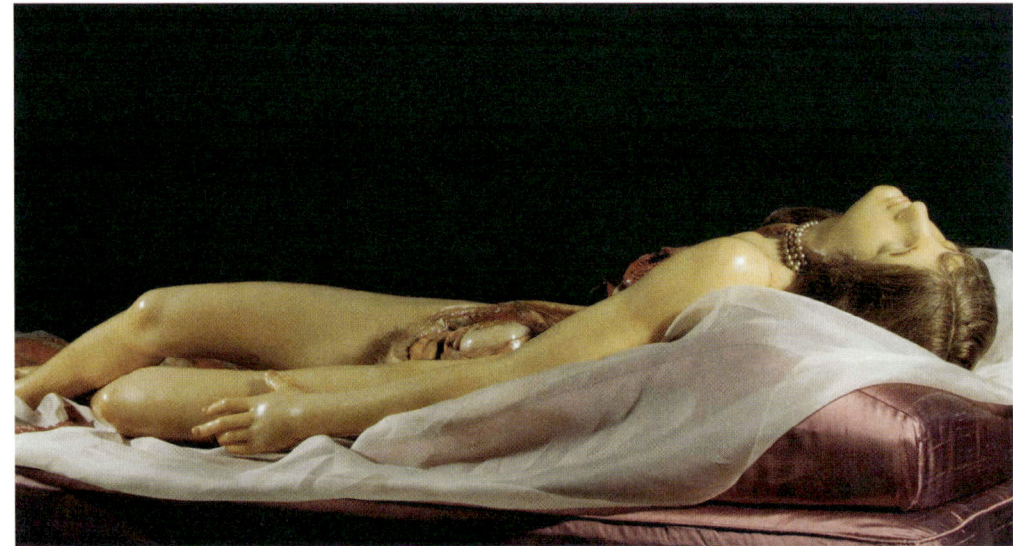
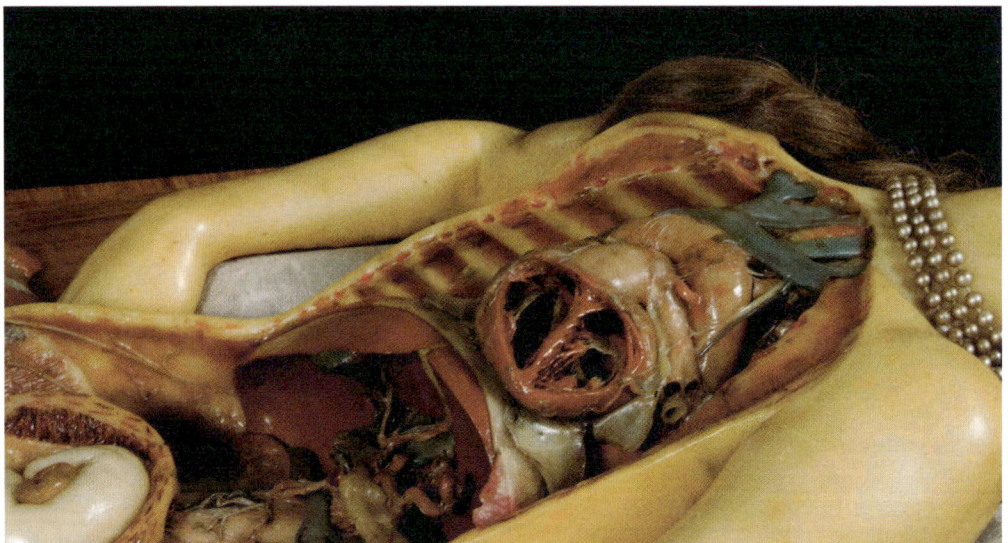
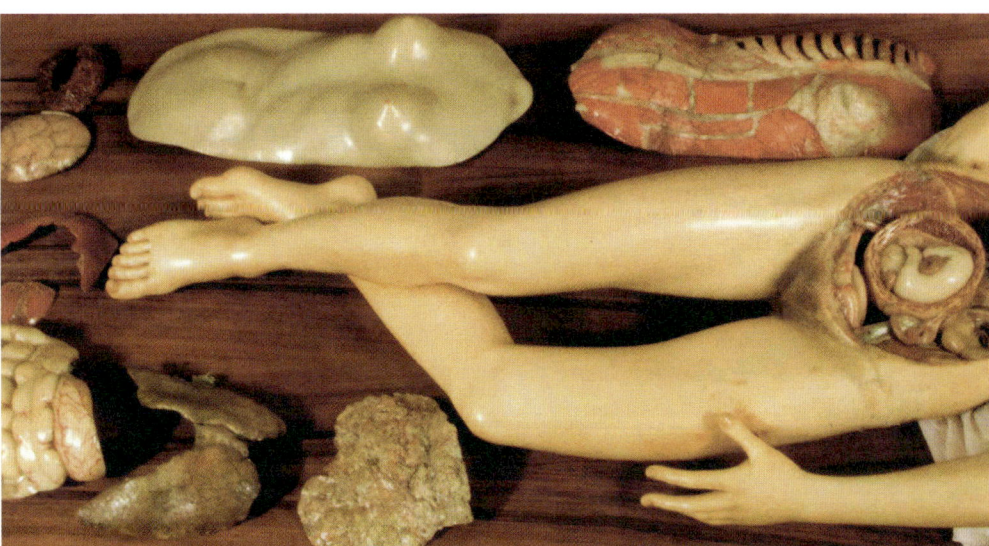

THIS PAGE & OPPOSITE *Details of Clemente Susini's Venerina. A fetus can be seen in the womb, although the model exhibits no outward signs of pregnancy.*

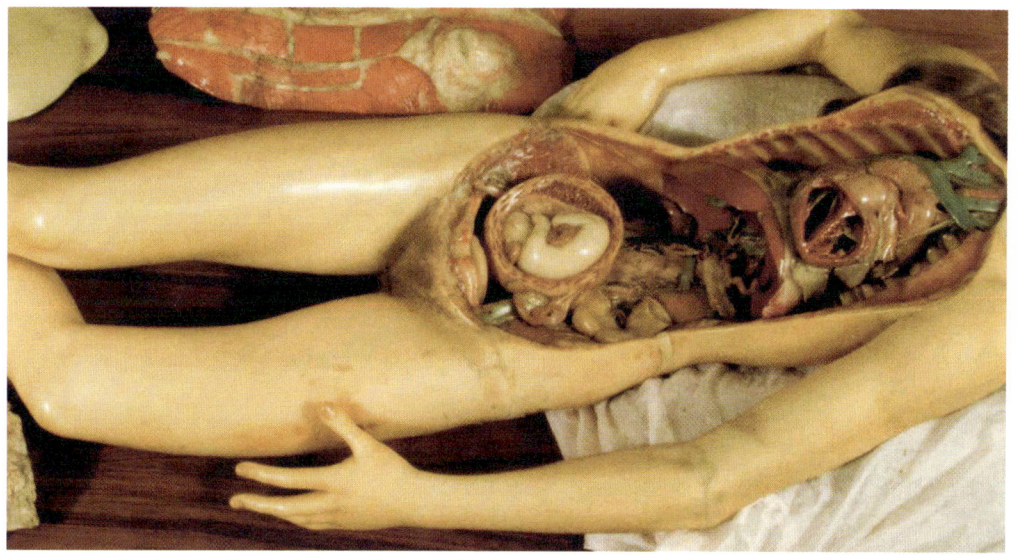
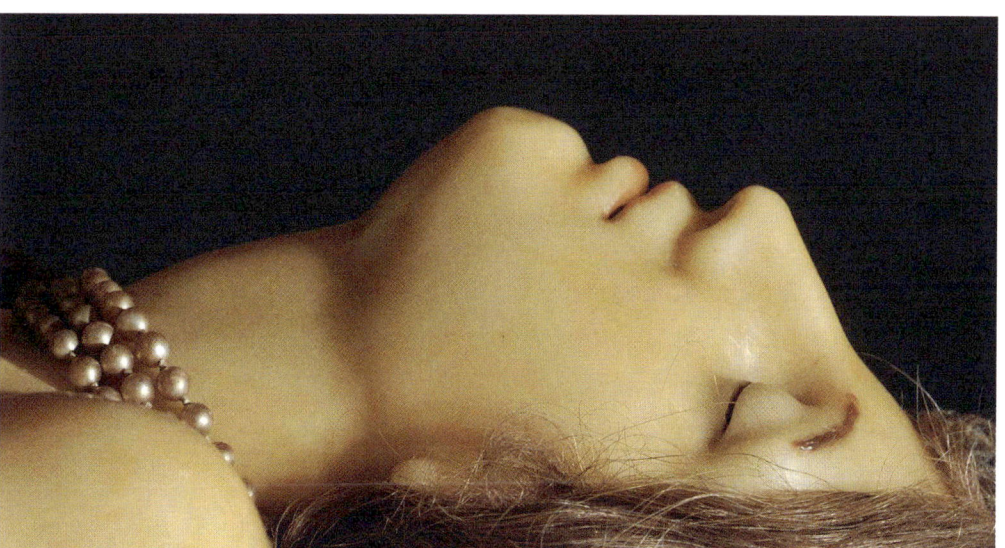
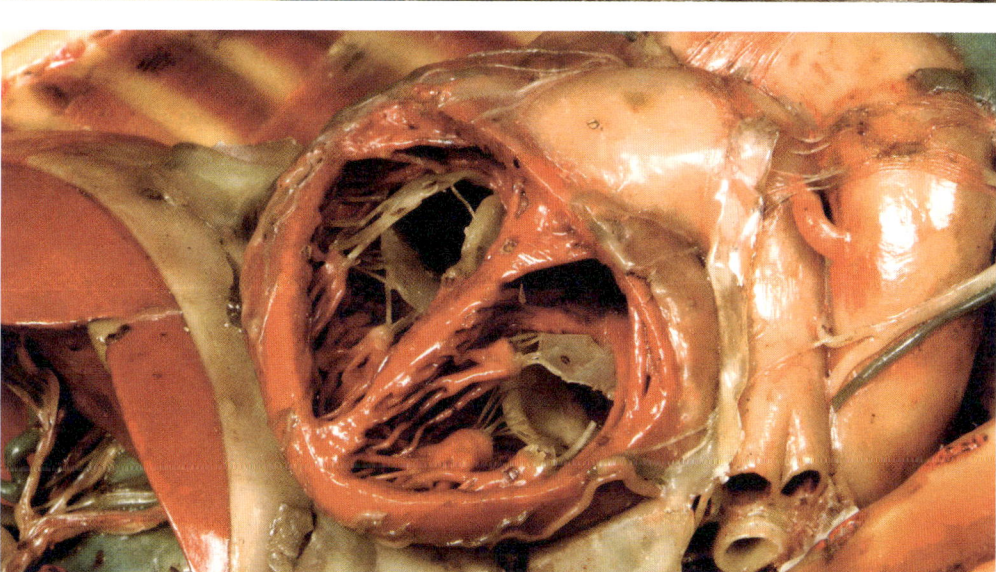

[2] *The Venerina may have been partly modelled on a young woman who died of a congenital heart defect. The left ventricle wall of a normal heart is three times thicker than the right, while here the walls are of equal thickness.* (115)

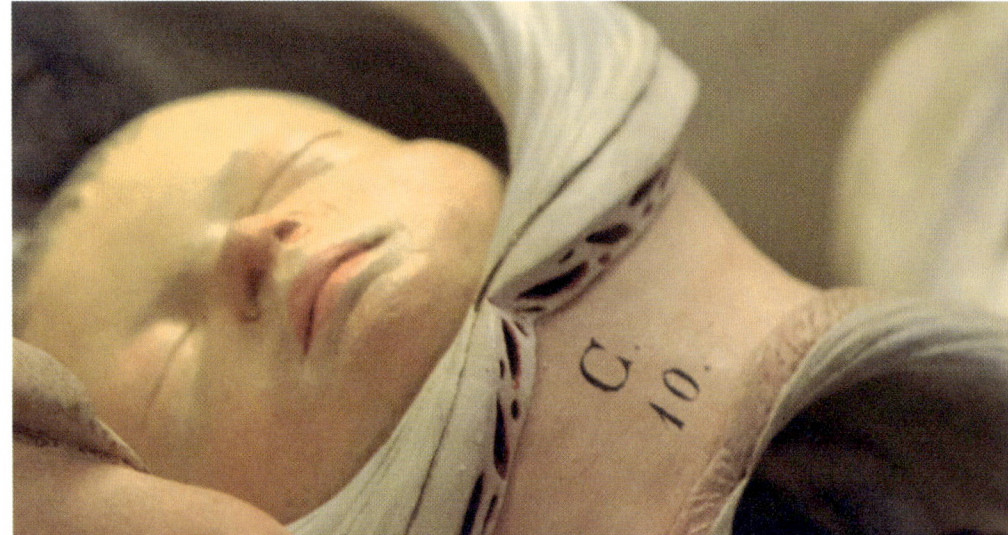
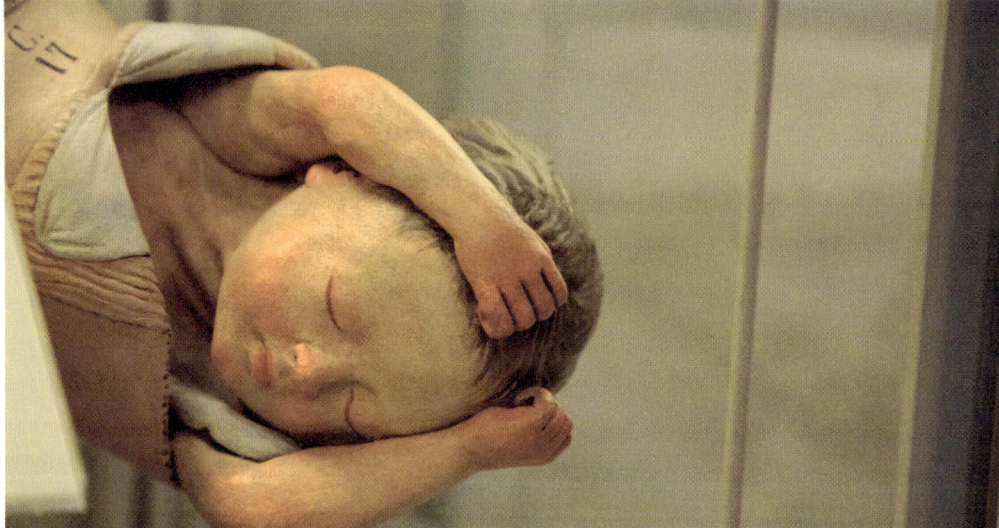
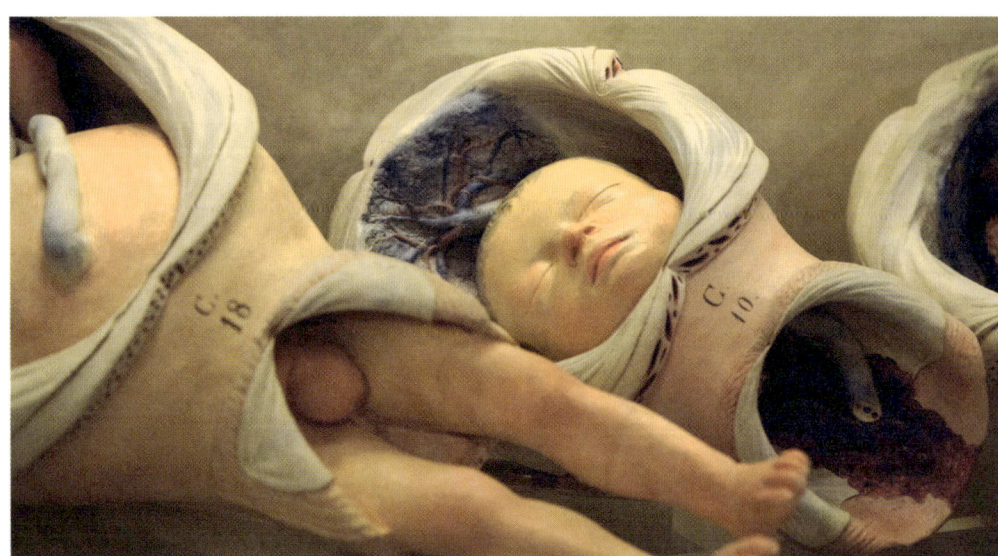

THIS PAGE & OPPOSITE Painted plaster 'obstetric phantom' models used for training surgeons and midwives. Made in the mid-eighteenth century by Giovanni Battista Sandri for the University of Bologna.

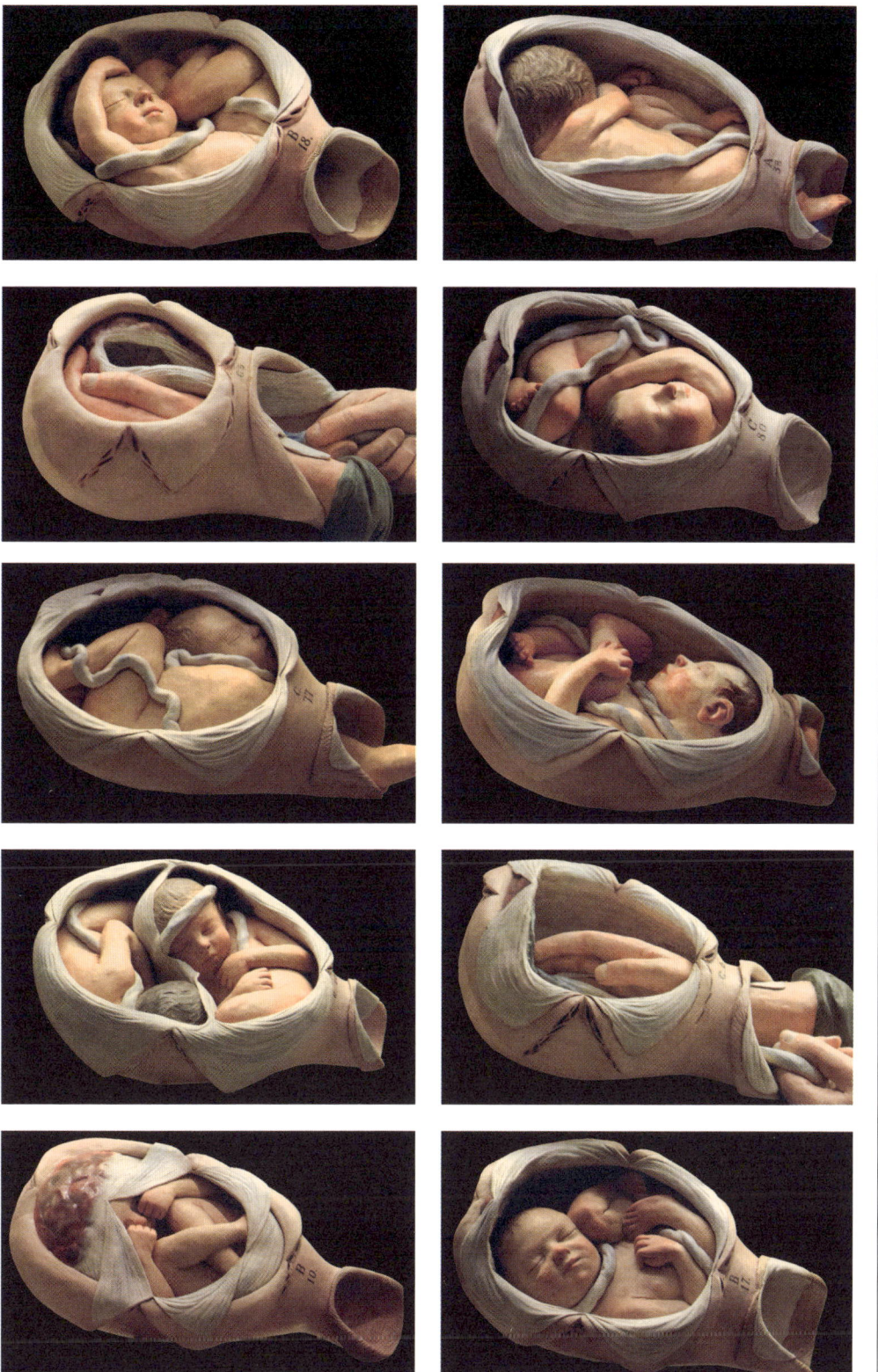

OVERLEAF *Poster advertising Chemisé's anatomical museum, showing eclectic displays, including anatomical and ethnographic models and men gathered around an Anatomical Venus (c. 1850).*

CHAPTER THREE

VENUS·AT·THE FAIRGROUND

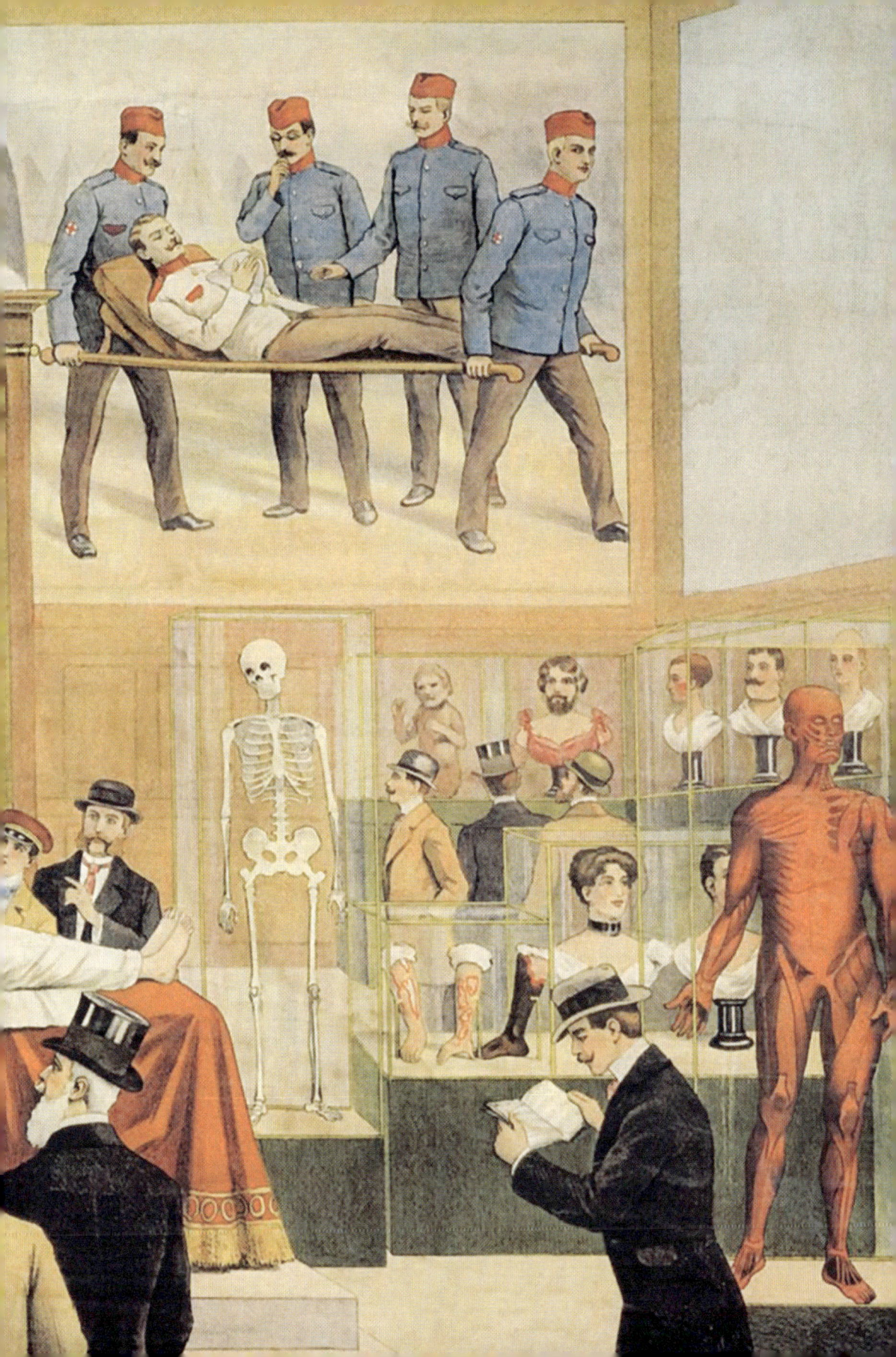

PREVIOUS
Poster by Adolph Friedländer showing the interior of a popular anatomical museum in Hamburg, Germany (1913). A gentleman has removed the heart of an Anatomical Venus before a crowd of captivated male onlookers.

O ur ancestors had very different ideas about viewing dead, dying, and anatomized bodies. Up until the early twentieth century, there were a variety of collectively organized rituals and amusements involving cadavers, characterized by a combination of the educational and the spectacular that can be shocking to contemporary sensibilities. They are a reminder that contemporary Western customs regarding dead bodies are far from universal. Before death was something that happened behind the drawn curtains of a bed on a hospital ward, when a 'good death' routinely included the open display of the corpse at home, and when life expectancy was far shorter and death was an occurrence more frequent in everyday life, familiar attitudes towards this most certain of events routinely included humour, pleasure, and undisguised fascination.

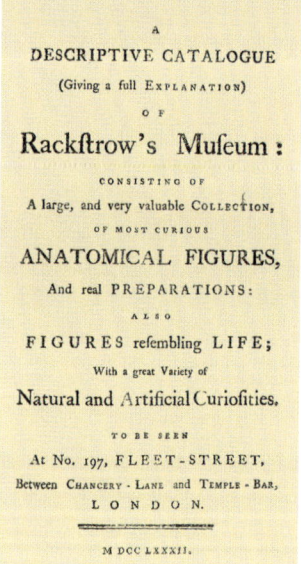

fig. 52

fig. 53

fig. 52 Miniature illustration showing an autopsy at Bologna's anatomical theatre in the eighteenth century. Public dissections took place here as part of the annual carnival festivities.

The Anatomical Venus was just one of many representations that played on this familiarity with death, with an added undercurrent of passive eroticism in the inert female body. As such, she expressed a fascination with the relationship between life and death, playing on voyeurism, desire, and possession. The scientific study of anatomy provided a safe, legitimizing frame through which the naked female body could be viewed, at a time when most displays of nudity were highly regulated. As medical historian Michael Sappol comments, 'Public discussion of sexual desire, ostensibly part of the medical technology of self-regulation, provided an opening through which the pleasure principle could be smuggled in, a fact that popular anatomists were well aware of, and manipulated to their benefit.'

Human dissection, despite its stench and gore, was, from the sixteenth century on, commonly performed for the general public at purpose-built anatomical theatres in university cities such as Padua, Pavia, Leiden, and Bologna. The theatre at the University of Padua, built in 1594, is the oldest known example that can be visited today. Anatomical theatres also served as early anatomical museums, exhibiting preserved specimens for the public all year round. Public dissections were sometimes even held—as in Bologna—as part of the pre-Lent Carnival (*carne vale*, 'farewell to meat') festivities. Carnival was seen as an appropriate time for such *utilia spectacula*, or 'useful spectacles', as bodies decayed relatively slowly in spring, and the general population was available to attend because they were already on holiday. The subject of the carnival dissection was, ideally, an executed criminal, but if no convict was available, another body would be found. These public events would be attended by a mix of students, aristocrats, and masked revellers.

fig. 53 Title page from the 1782 catalogue for Benjamin Rackstrow's museum, which was famous for Abraham Chovet's 'exceeding curious figure'—a waxwork model of a dissected pregnant woman whose glass veins flowed with red wine (see page 37).

fig. 54

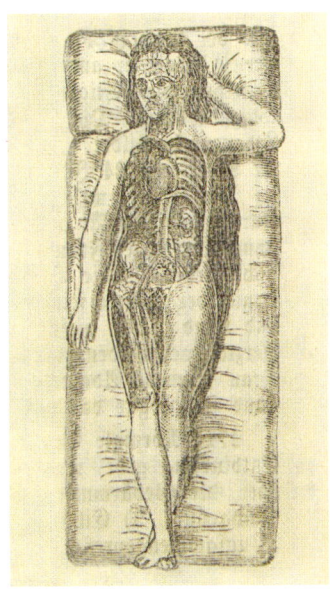

fig. 55

Carnival dissections were not just about education and entertainment; they also acted as a sort of morality tale, a secular, state-organized public ritual in which the disorder of sin was transformed, via the sacrifice of an executed criminal, into a social good. Another way of looking at these public dissections is as an expression of the Russian philosopher and critic Mikhail Bakhtin's notion of the 'Carnivalesque'; festivals invite participants to turn the everyday world on its head, subvert the status quo, blur class distinctions, and indulge in any number of usually unacceptable behaviours. Catering to an unruly audience of masked revellers who, as one visitor complained, 'instead of listening and learning created an uproar and conducted themselves "without due modesty"', the public dissection served to overthrow the rudimentary taboo around violation of the dead body, as well as providing a safe environment for the collective exploration of various anxieties about death and social regulation.

fig. 54 Cupid Taking Down the Smock of Venus, Louis-Marin Bonnet.

fig. 55 Illustration of a dissected Anatomical Venus from a nineteenth- or early-twentieth-century catalogue to Gassner's Anatomical and Pathological Museum, St Petersburg, Russia (see pages 127 and 136).

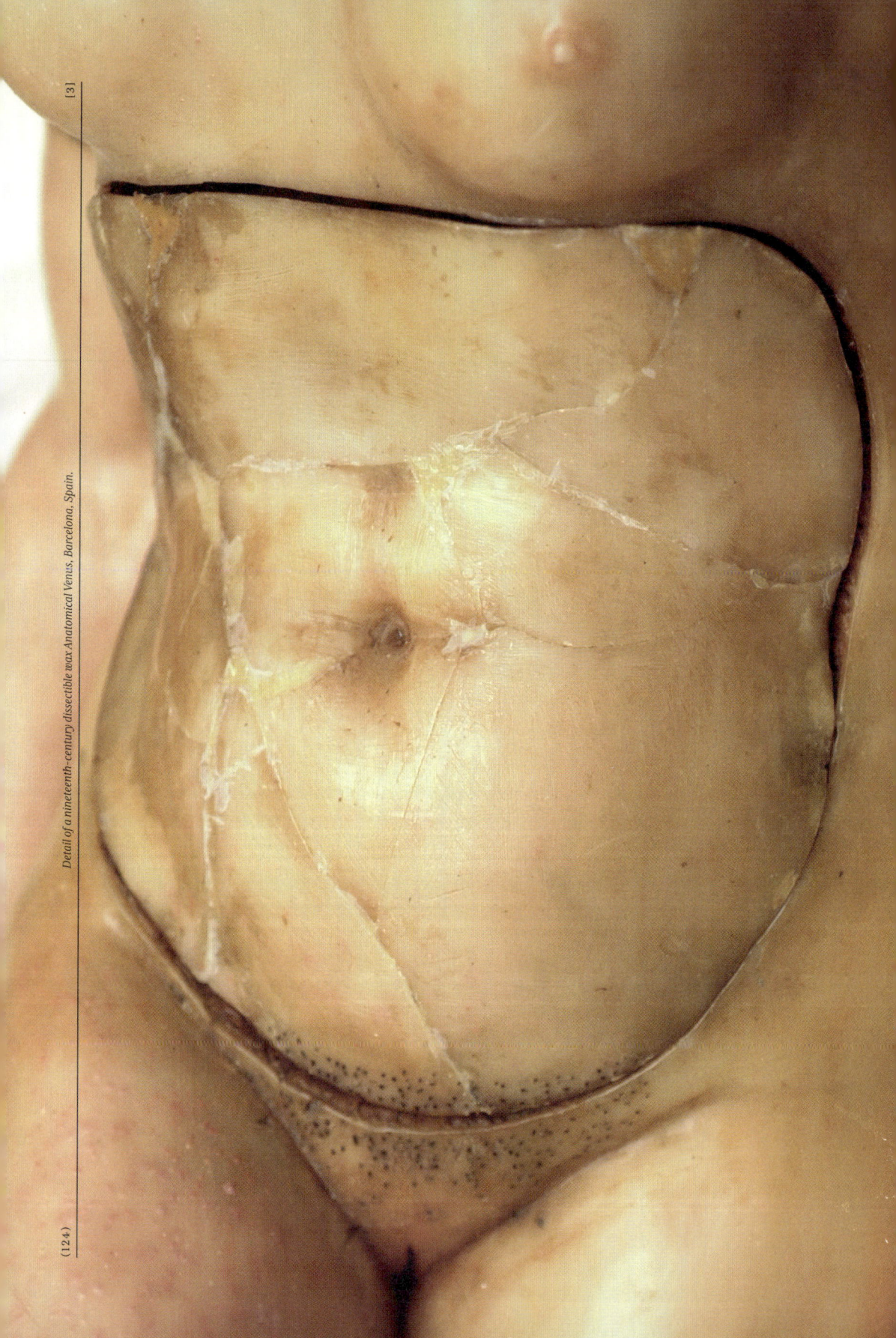

Detail of a nineteenth-century dissectible wax Anatomical Venus, Barcelona, Spain.

UNDER THE PRETENCE OF IMPARTING ANATOMICAL KNOWLEDGE, THIS FILTHY FRENCH FIGURE … IS EXHIBITED. IT IS A LARGE, DISGUSTING DOLL, THE ALVUS OF WHICH BEING TAKEN OFF LIKE A POT-LID, SHOWS THE INTERNAL PARTS, HEART, LIVER, LUNGS, KIDNEYS, &C AS REMOTELY FROM ANATOMIC PRECISION OR UTILITY AS ANY OF THE SIXPENNY WOODEN DOLLS WHICH YOU MAY BUY AT BARTHOLOMEW FAIR …. THE THING IS A SILLY IMPOSTURE, AND AS INDECENT AS IT IS WRETCHED.

Description of the 'Florentine Venus', which was exhibited in various London venues from 1825. From the British Literary Gazette. ('Alvus' is an obsolete term used here to refer to the abdomen.)

32 Internat. Handels-Panoptikum und Museum, München.

NB. In den Sälen der III. Etage befindet sich das

große anatomische Museum

mit über 700 Präparaten, dessen Besichtigung angelegentlichst empfohlen wird. Es haben in dasselbe nur Erwachsene Zutritt.

Jeden Freitag Nachm. 2 Uhr ab sind diese Sammlungen nur für Damen geöffnet.

Der Beginn der Vorstellungen auf den

Spezialitäten-Bühnen

in der I. und II. Etage, auf welchen zeitweise lebende Spezialitäten, Menschenragruppen, Abnormitäten, Illusionen etc. vorgestellt, und wissenschaftliche Demonstrationen gehalten werden, wird jedesmal durch Glockenzeichen in sämmtlichen Etagen bekannt gegeben.

Nachdem nunmehr der Besucher des Panoptikums am Ziele seiner Wanderung angelangt ist, wird sich in ihm auch gewiß das Bedürfniß regen, eine Erfrischung zu nehmen. Es sei daher der Besuch des im Parterre gelegenen, freundlich ausgestatteten

Panoptikum-Restaurants,

in welchem ausgezeichnetes Hackerbräubier, vorzügliche kalte und warme Speisen aller Art, Kaffee, Weine ic. ic. zu civilen Preisen verabreicht werden, bestens empfohlen. In demselben finden täglich Concerte statt.

Frühschoppen-Concerte von 10—1 Uhr
Abends 8—11 „
Sonn- und Feiertags-Concerte Nachmittags von 4—6 Uhr.

Mit Hoher Obrigkeitlicher Bewilligung.

Der Italiener F. Dominikini,
welcher mit einem zahlreichen, aus 50 Wachsfiguren bestehenden Kunst-Kabinet, welches auf seiner Reise durch ganz Deutschland, Moskau und mehreren Hauptstädten Rußlands mit großem Beifall aufgenommen wurde, in Riga angekommen, ist vom Willens, dem hiesigen hochzuverehrenden Publiko seine

Wachsfiguren in Lebensgröße

MUSEO ROCA
INSTALADO EN LA FERIA
Bajo el Control de la Dirección General de Sanidad

El primer **MUSEO DE CERA** de España
presenta en forma real y para educación del pueblo

Los Estragos del Barrio Chino
La degeneración del hombre por el vicio.
Las grandes plagas sociales.
ESTUPEFACIENTES
Morfina - Opio - Cocaína - Éter - Alcohol.
Sus desconsoladores efectos. — El vicio al descubierto.

NOVEDAD SENSACIONAL
EL HOMBRE MONO
En tamaño natural y auténtico.
Como venimos al mundo.

El único ejemplar existente en España de la

Araña Gigante del Japón
lo presenta el MUSEO ROCA.

Más de 500 ejemplares en cera.
Fetos humanos auténticos.

Las Hermanas Siamesas
ya conocidas por la prensa española.

Galería de curiosidades.
Galería de hombres célebres.
Galería de monstruos humanos.

Este Museo cuenta con personal competente para dar explicaciones científicas al público que lo visite.

Si no visita el MUSEO ROCA adquiere cargo que no ha visto nada

ADVERTENCIAS IMPORTANTES:
- Solo se permitirá la entrada a los mayores de 18 años. Tiempo de permanencia ilimitado.
- Abstenganse personas impresionables.
- Centros culturales, mutilados de la guerra y entidades benéficas, entrada gratis visitándolo en colectividad y de acuerdo con la Dirección.
- Para dilucidar si la exhibición de cera representa en su fin científico y dilucidante moral, se espera de la cultura del público se abstenga de hacer cualquier manifestación que pueda distraer la atención del profundo estudio que ofrece este Museo.
- El espectador que una vez visitado el MUSEO ROCA, direse por mal entendido el importe de la entrada, tenga la bondad de reclamarlo y le será devuelto en el acto.

PROHIBIDO FUMAR.

Lea este programa y no lo tire, entréguelo a un amigo y se lo agradecera.

ANATOMICAL MUSEUM, CHESTNUT ST. No. 807

DRS. JORDAN & DAVIESON'S
MAGNIFICENT
Anatomical Museum
807 CHESTNUT ST.,
Opposite Continental Hotel, Philadelphia.

THE FINEST IN THE WORLD.
NONE TO EQUAL IT!!
SHOULD BE VISITED BY ALL.

INSTRUCTIVE,
AMUSING,
SCIENTIFIC,
AND ARTISTIC.

Acknowledged by the Medical Profession and the Press to be the Grandest Collection of Anatomy and Science ever exhibited.

The magnificent Venus, taken from life.
The Dying Soldier on the Field of Battle.

THE SLEEPING BEAUTY.
The Maniac of Sing-sing; The Guillotine Execution; Embryology, Midwifery and Pathology; Male and Female Skeletons; Birds, Beasts, Fishes and Reptiles, with skeletons of the same; The Strongest Magnet in the World.

COLLECTED AT AN ENORMOUS EXPENSE.

Thousands of Natural Preparations and Other Marvels,
ALTOGETHER FORMING A COMPLETE

PALACE OF WONDERS.

Open Daily, For Gentlemen Only.
807 Chestnut Street,
Opposite Continental Hotel, PHILADELPHIA.

Admission 50 cents.

J. MOORE & SONS, PRINTERS, 1135 & 1137 RANSOM ST.

KREUTZBERG'S
EUROPEAN ANATOMICAL
AND
HISTORICAL MUSEUM
Occupying the whole Building 729 Chestnut Street, between 7th & 8th.

I.
THE SPANISH INQUISITION
OR THE TORTURE OF THE MIDDLE AGE

Consists of Life-Size Figures, in Wax, representing the Tortures to which Condemned Prisoners were subjected.

THE IRON MAIDEN.

The Torture Boot, Thumb Screws, Foot Screws, The Torture Spider, The Girl on the Torture-Bench.

The tortures of the inquisition represented in full-length figures.

The Man on the Torture Rack, Leg Screws, and various objects too numerous to mention.

II.
Anatomical Museum of Science, Art and Nature.

The finest collection in the United States, comprising over Six Hundred different specimens of interest. Among the Articles especially noticeable, the following are named:

THE ANATOMICAL HERCULES,
One of the Greatest Artistical Works in America. On Exhibition for the First Time in this Country.

Venus and Amor
One of the Greatest Masters of Venice. On Exhibition for the First Time in this Country.

THE SIAMESE TWINS, LIFE SIZE. THE TWO Nightingales!

Cora Pearl! The Handsomest Woman of France. On Exhibition for the first time in this Country.

THE TATTOOED MAN OF BURMAH. Who escaped the Greatest Execution during the Vienna Exposition.

PRINCE BISMARCK! LIFE SIZE. **NAPOLEON III.**

YOUNG ARABIAN GIRL!
THE DIFFERENT NATIONS OF ASIA, AFRICA AND AUSTRALIA.

THE GREAT CAESARIAN AND CIRCUMCISION OPERATIONS!
2 Natural, the Pride by L. PEREZ. **HYDROCEPHALUS!** (FEMALE) born by Richardson. She is the only one of the kind in the Country.

EGYPTIAN MUMMY!
CORPSE OF A WIFE OF NATURAL SIZE!

DREADFUL EFFECTS OF TIGHT LACING! Exhibiting Model of a Young Lady 18 years of age, who had a had in consequence of the pernicious habit.

ANATOMY of the EYES and the EAR. Curious and Interesting.

WONDERFUL MALFORMATION IN CHILDREN.
Two-Headed, Four-Armed & Double-Bodied

The Dying Zouave, The Senegambian Chief

Representing the last parting movement after a hard fought in the Light Armies.

A wonderfully preserved body, prepared in a manner unknown, in the Annals of Embalming Science.

Cleopatra & her Family.
The greatest artistical work here made in wax; there is no second one of its kind in the World.
Specially Made for the Centennial!

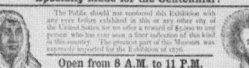

The Public should rest convinced that this Exhibition will in any case before exhibited in this or any other city of the United States, nor are the it a reward of praise for any person who has ever seen a first exhibition of this kind in this country. The greatest part of the Museum was especially imported for the Exhibition of 1876.

**Open from 8 A.M. to 11 P.M.
FOR GENTLEMEN ONLY!**

ADMISSION 50 CENTS.
Open from 8 A.M. to 11 P.M.

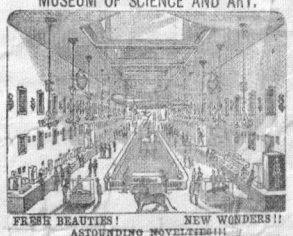

A selection of advertisements for displays of waxworks at museums and travelling shows as far afield as Latvia (opposite, top right), Philadelphia, USA (opposite, bottom right), Barcelona, Spain (this page, top left), and St Petersburg, Russia (this page, bottom right). The line between scientific museum exhibit and fairground entertainment is often blurred. Barcelona's Museo Roca (opposite, bottom left), is said to have been 'installed at the fair under the control of the General Directorate of Health'; it features a show on 'the degeneration of man through vice' alongside the 'new sensation' known as The Ape Man. Drs. Jordan and Davieson's establishment at 807 Chestnut Street, Philadelphia is variously advertised as a 'Magnificent Anatomical Museum', 'For Gentlemen Only' (opposite, bottom centre) and a 'Gallery of Anatomy and Museum of Science and Art' (this page, bottom left). In 1894, visitors to Handel's Panopticon and Museum in Munich, Germany (opposite, top left) could enjoy both the anatomical exhibition on the third floor and the restaurant downstairs, where concerts were held. Many of the museums' proprietors had backgrounds in showmanship; Sanger's Metropolitan Collection of Animated Wax Works and Alabaster Models (this page, top right) was run by 'Lord' George Sanger (1825–1911), a hugely successful showman who also founded Astley's Amphitheatre in London, UK.

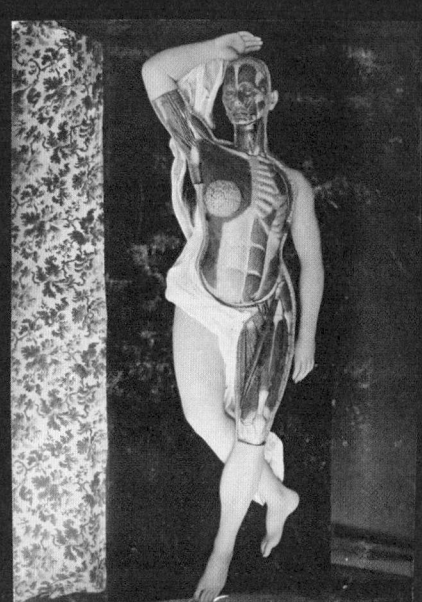

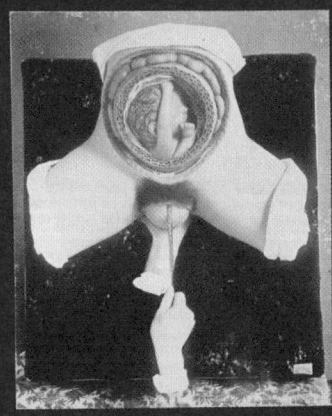

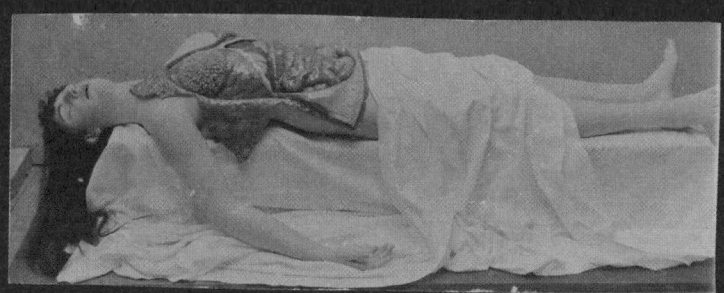

Early photography of wax anatomical models created in the workshop of Rudolf Pohl (1890–1910). The models were intended for exhibition in panopticons, and include 'Natural childbirth of an Indian woman' (opposite, top left); the effects of corsetry (this page, below), a staple of such museums; a dissectible Anatomical Venus, shown dissected (opposite, top right) and undissected (this page, centre left); a collapsible Anatomical Venus (opposite, bottom); an operation for gastric cancer (this page, top left); representations of breech births and normal births; and an operation on a grey cataract (opposite, centre left above).

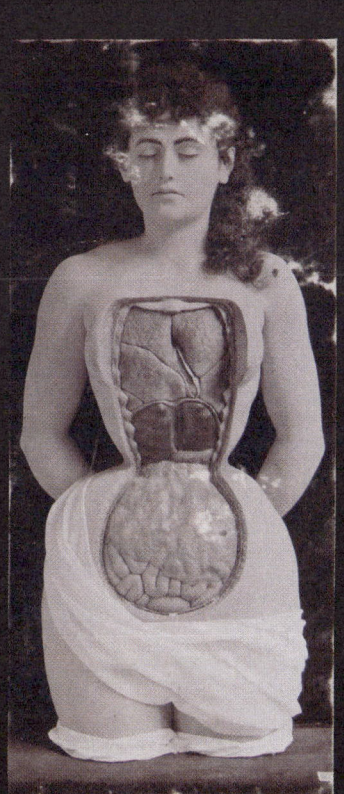
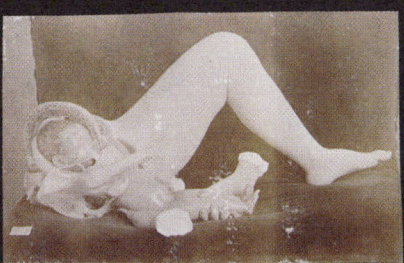

A similarly carnivalesque 'spectacle of death' could be enjoyed at London's Tyburn gallows (now the site of Marble Arch), where public executions were a popular entertainment that drew picknicking families among other onlookers. In 1783, when hangings were moved behind the closed doors of Newgate Prison and were no longer open to the public, the Newgate Calendar continued to publish vivid details, including images of waxwork models and tableaux of notorious felons that had been displayed at fairgrounds, maintaining the mixture of thrill and moral fable. Particularly vivid crimes were sometimes memorialized in what were called murder pamphlets, the precursors of true crime novels. These cheap printed works contained what was claimed to be a true account of each murder, often including the narrative along with the trial transcript or written confession of the murderer before his or her execution.

The Anatomical Venus was one of the most popular and lasting spectacles involving the simulation of a dead or anatomized body, one that carefully distanced itself from the gore and stench of actual dissection. The use of such models

fig. 56 Emil Eduard Hammer's The Nightmare, a tableau featuring a voluptuously swooning, life-sized wax woman, which was inspired by Fuseli's painting of the same name (see fig. 57). It was exhibited at Hammer's popular Munich panopticon.

fig. 56

fig. 57

fig. 57 Henry Fuseli's The Nightmare (1781). The painting was widely reproduced in Fuseli's lifetime.

was increasingly viewed as a possible solution to the nineteenth century's proliferation of criminal grave robbers and 'resurrection men' who stole human corpses—most often those of the poor or marginalized—to sell in the lucrative cadaver trade to medical schools, which needed a constant supply for students to dissect. This huge demand for corpses culminated in a scandal in Edinburgh in 1828, when Irishmen William Burke and William Hare murdered sixteen people to sell their bodies for dissection to Scottish anatomist Dr Robert Knox. In 1832, the Anatomy Act granted wider licences to dissect donated bodies but such scandals had made dissection increasingly distasteful to the public. With their impeccably artful appearances of being comfortably asleep, Anatomical Venuses, Eves, and Adonises avoided the major barrier to popular anatomical instruction: the disgust generally associated with dead bodies. By 1825 and until 1900, London was always home to at least one Anatomical Venus. Often referred to as the 'Florentine' or 'Parisian' Venus—capitalizing on the suggestive connotations of continental Europe—her classically beautiful exterior and abject innards created a heady frisson that made her the popular centrepiece and signature attraction of a variety of amusements. The Anatomical Venus's display of anatomy provided a legitimizing frame through which the naked body—otherwise largely taboo—could be viewed.

The early- to mid-nineteenth century was a time of great change. In Europe and the United States, unprecedented numbers of people were migrating from the countryside to the city, leaving behind close-knit farming communities to enter a new urban industrial labour force. Members of the working class and burgeoning middle class had more free time and expendable income to pursue a variety of amusements. Edifying and respectable amusements—sometimes called 'rational amusements'—were particularly popular, especially those that were scientific, anthropological, or anatomical in nature. There was also great interest in the mechanical, due to a new intimacy with the world of machines; metaphors of bodies as machines abounded. A variety of attractions and

fig. 58 Guide to the Swedish Panopticon in Stockholm (1903), advertising exhibits depicting the royal family, adventurers, and criminals. The panopticon had three floors, and was open from 1889 to 1924.

fig. 59 Guide to Castan's Panopticum in Berlin, Germany (1879). Open from 1869 to 1922, Castan's was one of the most popular panopticons of its day.

fig. 58

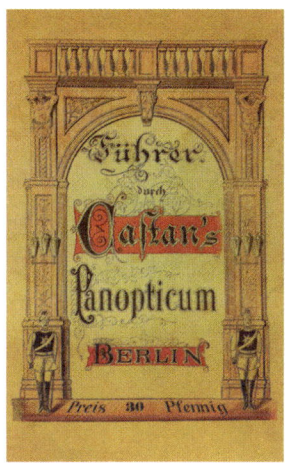

fig. 59

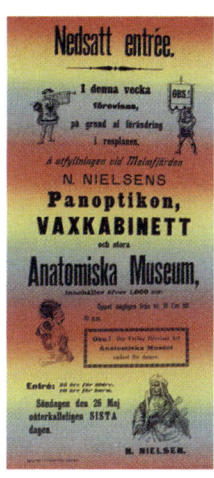

fig. 60

fig. 61

novelties emerged in answer to this demand, such as zoological and botanical gardens, taxidermy displays, projected microscope displays, and aquariums. There were also public exhibitions of automatons and other mechanical marvels; live displays of mesmerism, demonstrating Anton Mesmer's theories of animal magnetism; and even theatrical presentations of hypnotized hysterics, at 'father of French neurology' Jean-Martin Charcot's weekly open houses at his famous Salpêtrière hysteria clinic in Paris. Museums and exhibitions for a popular audience were a large and vibrant part of this new amusement scene.

One such leisure attraction was the 'panopticon'. In the panopticon, as critic Walter Benjamin (1892–1940) wrote, 'not only does one see everything, but one sees it all ways'. Entertaining and shocking, but also titillating, sensational, and educational, the panopticon was, like the United States' dime museums—such as P. T. Barnum's American Museum—a display of everything under one roof, focusing on the exotic and with an air of scientific objectivity. Such exhibitions

fig. 60 Advertisement for reduced entry to N. Nielsen's Panoptikon, Waxworks Exhibition and Large Anatomical Museum. Containing over 1,000 waxworks, it was open to ladies only on Fridays.

fig. 61 Guide to the Vienna City Panopticon, run by Louis Veltée. In 1896, Veltée began showing films at his establishment. He became known as 'the father of the Austrian movie theatre'.

THIS PAGE & OPPOSITE Wax models of two women demonstrating the external and internal effects of tight corsetry. From Castan's Panopticum in Berlin, Germany, which was open from 1869 to 1922.

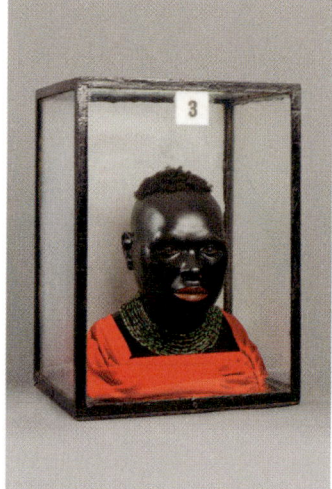

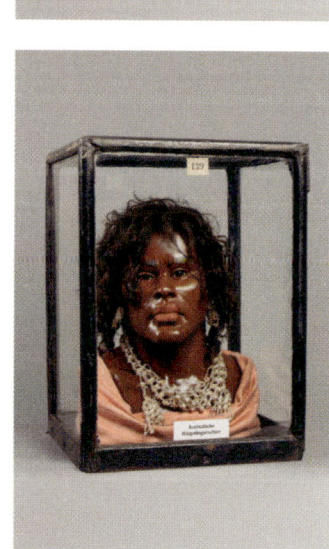

OPPOSITE *A series of wax ethnographic busts from Castan's Panopticum in Berlin, Germany (1869–1922). Such busts, as well as live 'exotic' people, were staples of the panopticon. At the height of its popularity, Castan's could attract as many as 5,000 visitors on a single Sunday.*

ABOVE *A wax bust depicting a woman suffering from a 'Feuermal', naevus flammeus, or 'port-wine stain' birthmark, in which swollen blood vessels create a reddish-purplish discoloration of the skin. This half-afflicted, half-healthy presentation recalls the conventions of earlier memento-mori artworks, while her necklace evokes that worn by the Medici Venus.*

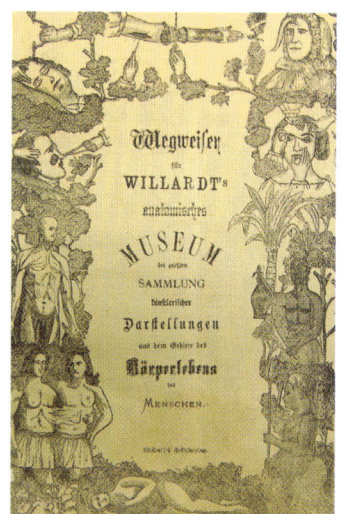
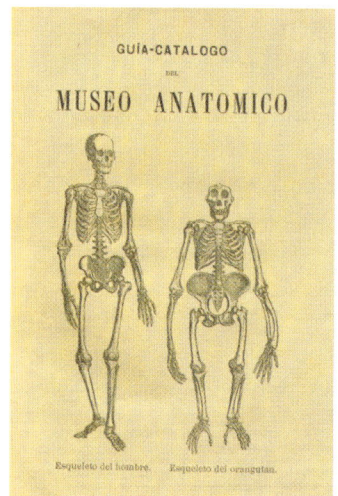
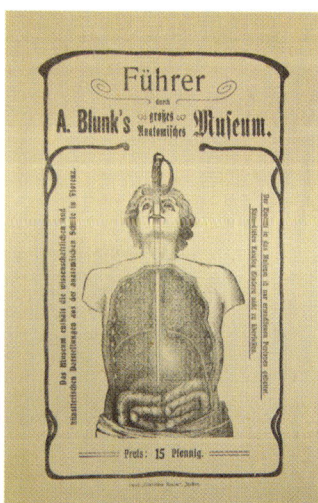

IT WAS A COLLECTION OF ANATOMICAL MODELS AND DISSECTIONS, WITH REPRESENTATIONS OF SKIN AND VENEREAL DISEASES, MOST IMPROPER FOR PUBLIC EXHIBITION, AND CALCULATED TO EXCITE THE MORBID CURIOSITY OF THE YOUNG TOGETHER WITH ITS PECULIAR FORMS OF HYPOCHONDRIA. VILE PAMPHLETS WERE ON HAND TO INDUCE THOSE HAVING OR FEARING DISEASE TO CONSULT THE PROPRIETOR. THE HARM WHICH THIS SINGLE ESTABLISHMENT MUST HAVE DONE CANNOT BE CALCULATED.

An excerpt from the editorial that appeared in the Boston Medical and Surgical Journal on 24 July 1873. It announced the closure of Dr Jourdain's Parisian Gallery of Anatomy 'in the cause of morality and public decency'.

Dieser Katalog annullirt alle vorhergehenden

ILLUSTRIRTER KATALOG
der
KUNSTANSTALT u. KUNSTHANDLUNG
EMIL ED. HAMMER, MÜNCHEN.

Nr. 54. Badende Nymphe, von Emil Ed. Hammer.

Preis dieses Kataloges Mk. 2.—. An Kunden gratis.

Nachdruck jeder Art verboten.

fall somewhere between aristocratic cabinets of curiosity and modern museums, displaying for a popular audience anatomical and pathological waxworks, human specimens, death masks of celebrities and murderers, ethnographic busts depicting the 'races of man', and assorted curiosities, including elephant tusks, mummies, stuffed alligators, and monkey skeletons. Panopticons also presented live acts, such as singers, dancers, ventriloquists, hunger artists (who 'starved' for an audience's entertainment), living 'freaks', and 'ethnic rarities'.

Popular anatomical museums shared many of the interests of the panopticon, but focused more narrowly on displays of human anatomy and pathology. Run for profit, these were close cousins of the institutional medical museums available to medics and students, such as that at London's Royal College of Surgeons. Popular medical museums exhibited real human specimens and a variety of models made of wax and other media in the name of public health, entertainment, and self-knowledge. Some had their own buildings, while others travelled the fairground circuits, setting up shop among other attractions. Most

fig. 62 Pamphlet advertising Signor Sarti's celebrated Florentine Anatomical Venus, published between 1847 and 1854. It particularly urged women, perceived as 'the nurse, trainer, and teacher of the whole human family', to attend the exhibition.

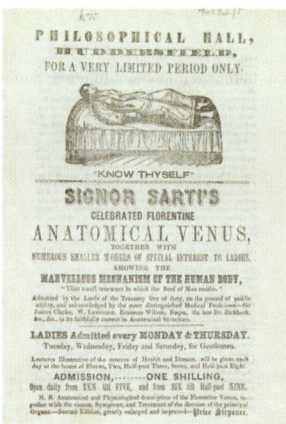
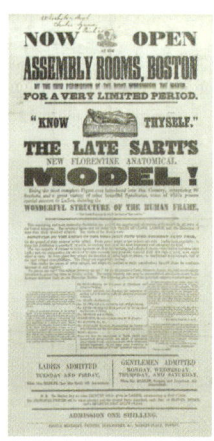

fig. 62   fig. 63   fig. 64   fig. 65

fig. 63 Advertisement for Antonio Sarti's Florentine Venus, to be displayed in Boston, UK. The model's internal organs and muscles could be taken apart, and were crafted to demonstrate different pathologies; for example, a model of the liver showed 'the effects produced by Intemperance and Excesses in Eating'. 'Know Thyself' was a phrase commonly associated with the showing of waxes.

major cities in Europe and the United States boasted at least one such establishment, if not many, and they were visited by a broad cross-section of society. Often, they would present a programme of educational lectures; sometimes these would be open to men only, but there were also special 'ladies nights' for women to learn about the mysteries of anatomy—important for their role as the family's caretakers—without the presence of gentlemen. Special attention was paid to the science of reproduction, including sexual anatomy, embryology, and sexual pathology.

The eclectic nature of the exhibits at such establishments can be gleaned by perusing the guidebooks: Liverpool's Museum of Anatomy, on tour in Blackpool in 1885, lists a 'full length Florentine model of Louise Lateau', stigmatic and prone to ecstatic trances; the head of a New Zealand chief killed in conflict with England; the skeleton of a viper; the head of a mummy; a recumbent Florentine Venus; the 'face of a bachelor, a confirmed onanist'; a model 'of the head and neck showing the awful and degraded state into which women come when they disobey the laws of God'; a model demonstrating the 'displacement of the womb by tight lacing'; and a model depicting 'circumcision as performed among the Jews'. Meanwhile, the European Anatomical, Pathological, and Ethnological Museum of Professor Charles Kreutzberg boasted in its 1875 catalogue 'the prolonged clitoris of a Hottentot'; models of diseases of the eye and uterus; models demonstrating the effects of syphilis and gonorrhoea; other models showing 'effects of masturbation on a young wife'; a 'corpse of a woman...dissectible in all its parts'; 'the head of a Hungarian...[with] a peculiar growth from the forehead'; a statue of Venus and Amor; a bust of the famous English courtesan known as Cora Pearl; and an Anatomical Adonis.

Sexual hygiene and the organs of generation were a strong focus of these displays, as were the erotic and exotic. At a time when nudity was considered unsuitable for display unless it was 'furnished with moral claims', the veneer of anatomical study provided just that. In addition to waxes depicting the male and female genitalia, there might also be tableaux depicting a gorilla ravishing

figs 64 & 65
1883 catalogue to Madame Tussaud & Sons Exhibition in London, UK. Marie Tussaud learned the craft of waxworking from Philippe Curtius (1737–94), a Swiss physician and master ceroplastician who was presented as her uncle but who some believe was her illegitimate father.

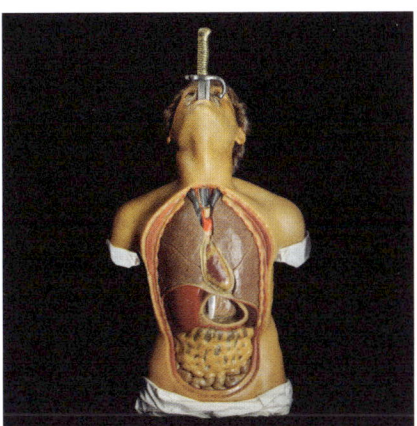

fig. 66

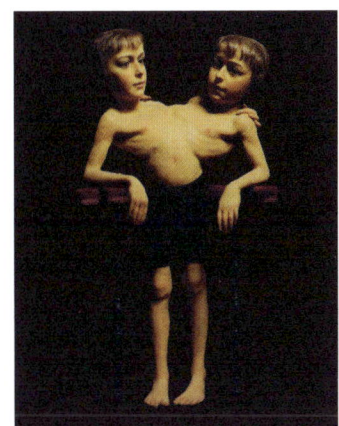

fig. 67

fig. 68

a pretty girl, odalisques in boxes, or a monstrous beast atop the torso of a swooning, bare-breasted woman, inspired by Fuseli's famous painting *The Nightmare* (1781) (see page 130).

Popular anatomical museums were also notorious for their terrifying displays of bodies—particularly genitals—ravaged by the 'moral diseases'. This was a time when syphilis was still fatal, mysterious, and somewhat ubiquitous. 'Spermatorrhea', believed to be caused by over-indulgence of the sexual appetite and onanism (masturbation), was considered a real threat and was also an extremely common diagnosis. *The People's Common Sense Medical Adviser in Plain English; or, Medicine Simplified* (1895) describes the symptoms of spermatorrhea as:

> ...loss of nervous energy, dullness of the mental faculties, and delight in obscene stories...the face bloated and pale, and the disposition is fretful and irritable; the appetite is capricious...pains in the chest, wakefulness, and during the night lascivious thoughts and desires.

fig. 66 Wax model (c. 1900) showing a man swallowing a sword, his internal organs bared. From the workshop of Rudolf Pohl, Dresden, Germany.

figs 67 & 68 Nineteenth-century, life-sized wax model of Siamese twins Baptisto and Giovanni Tocci. From Dr Pierre Spitzner's museum of anatomy and ethnology, Paris, France.

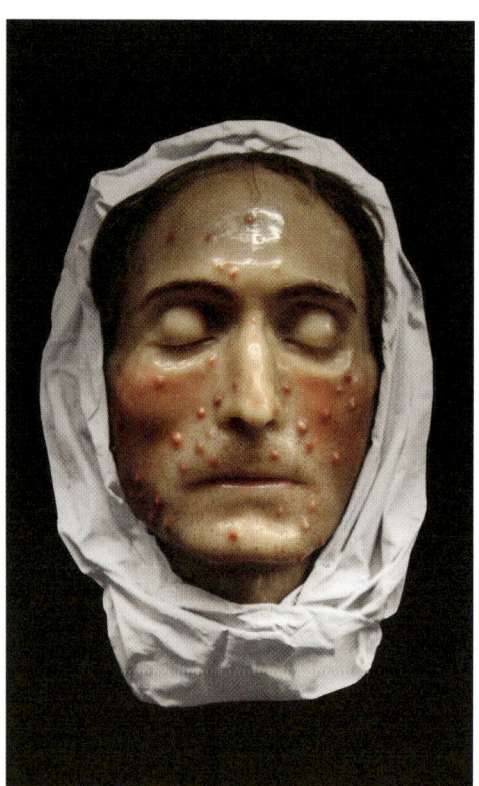
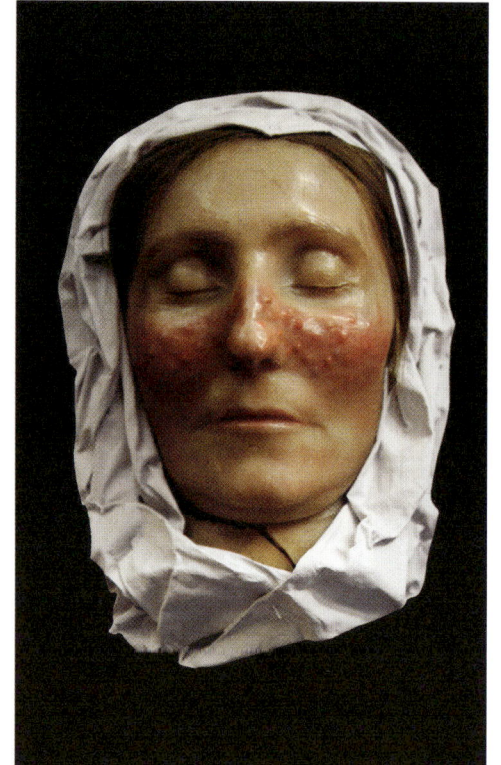
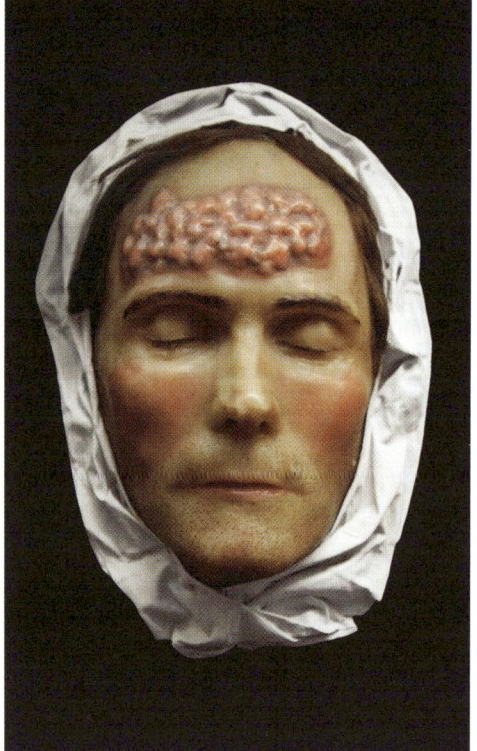
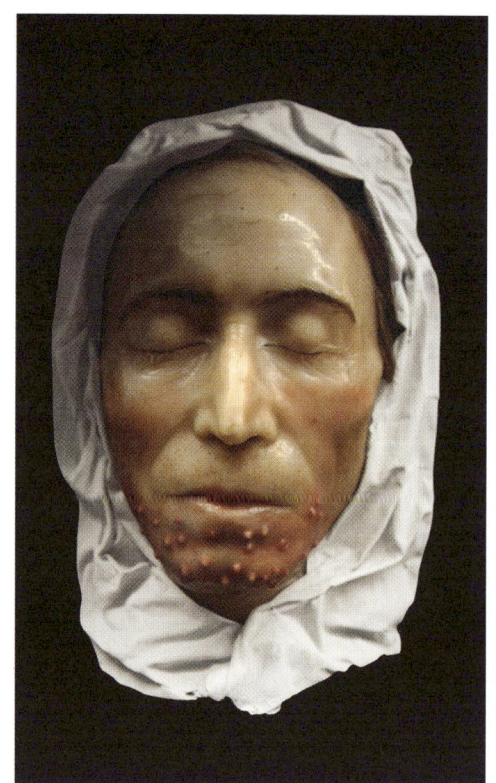

Late-nineteenth-century dermatological moulages from the Museo de la Medicina Mexicana, Mexico City. In replicating life so closely, the moulages also function as portraits of the afflicted.

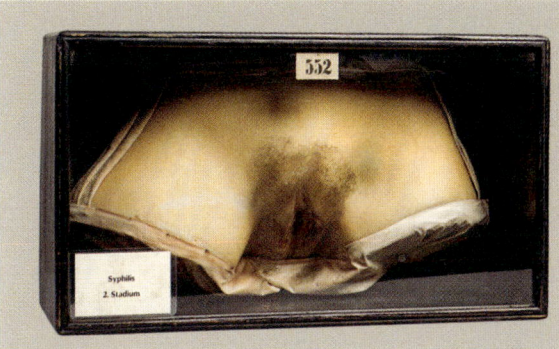
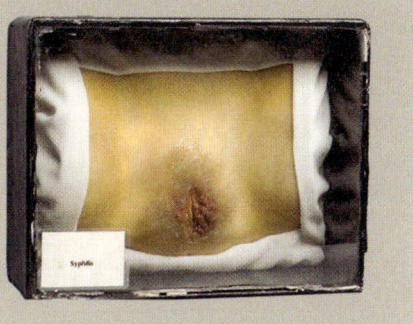
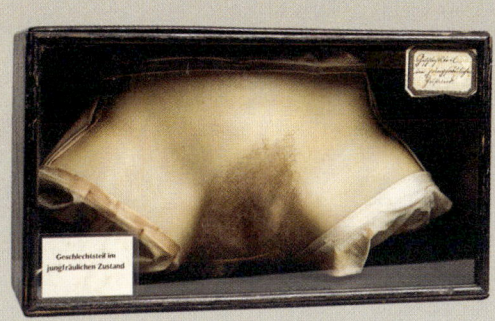
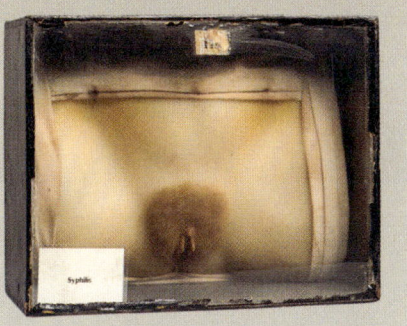
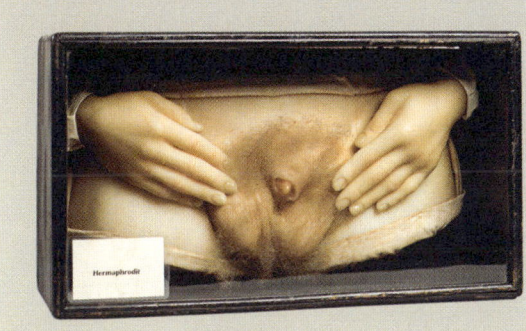
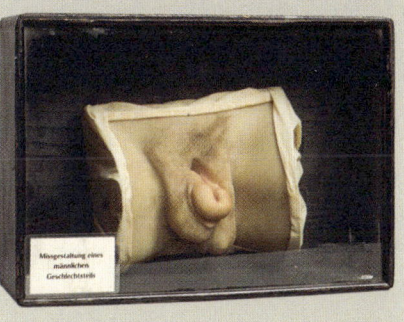
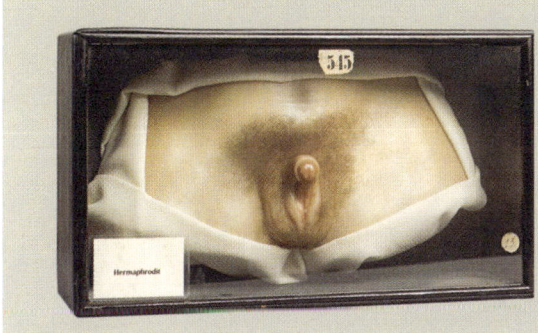
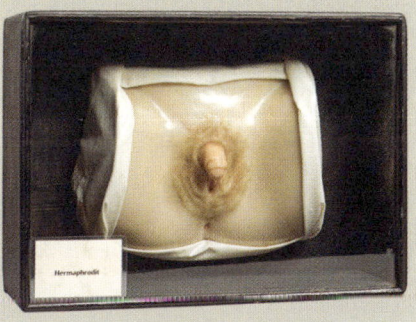

A collection of pathological genitals displayed at Castan's Panopticum, Berlin, in 1873. They show the effects of syphilis and other sexually transmitted diseases.

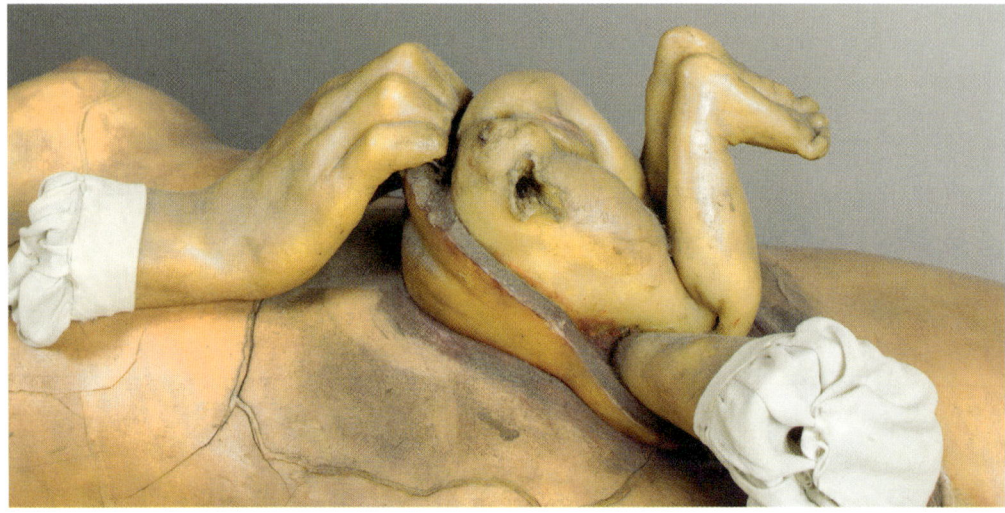

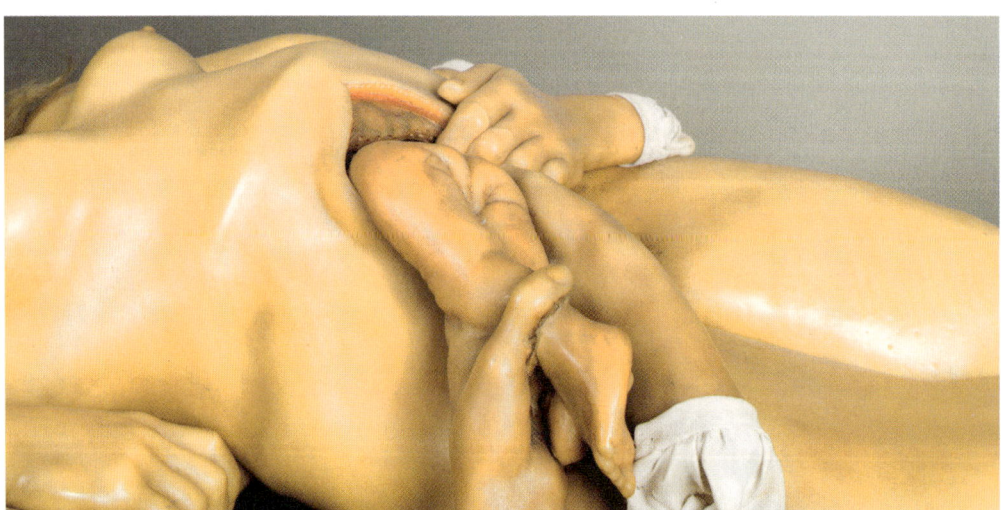

THIS PAGE & OPPOSITE Anatomical Venuses, once on view at Castan's Panopticum, Berlin, demonstrate caesarean sections and difficult natural births with the aid of phantom hands.

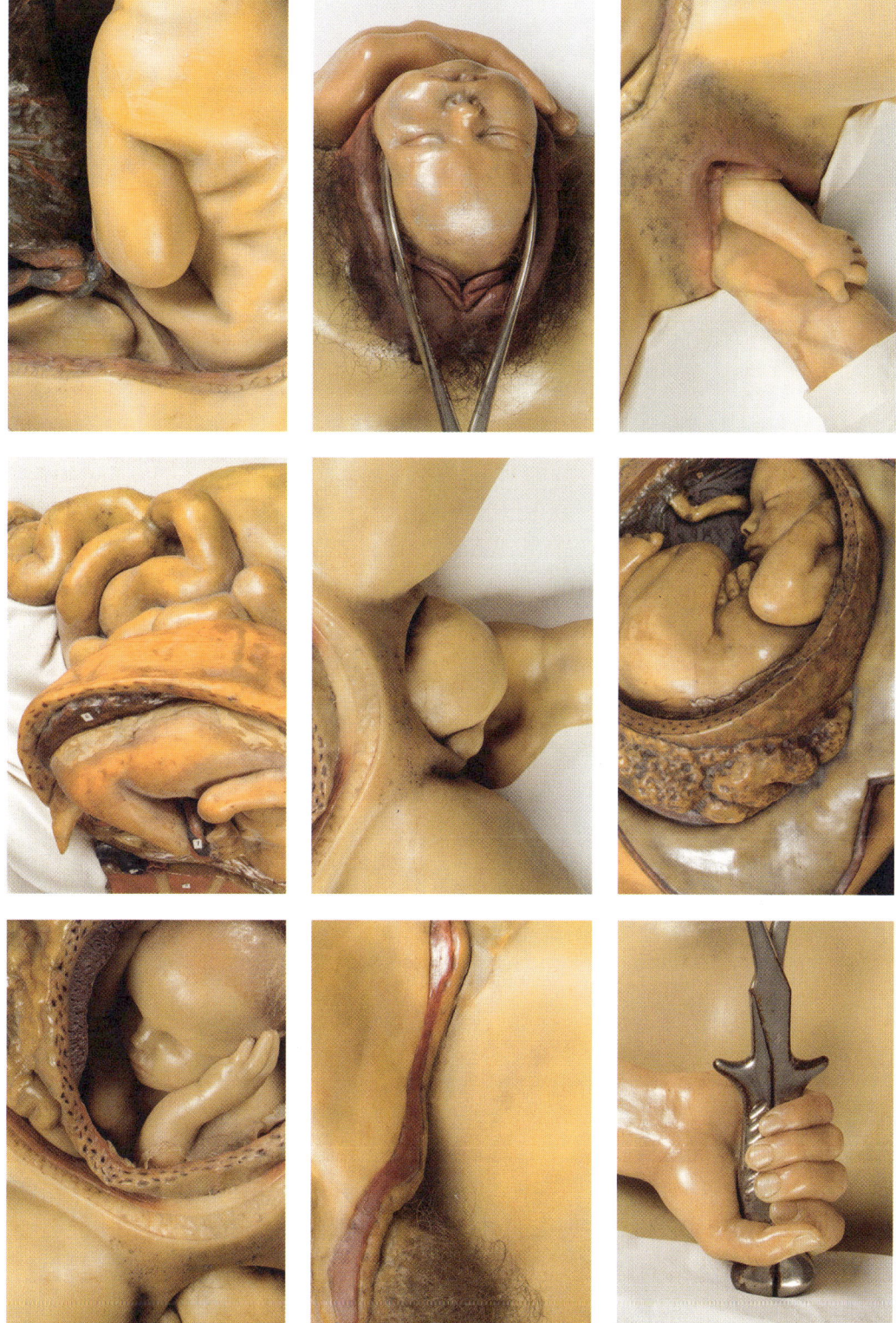

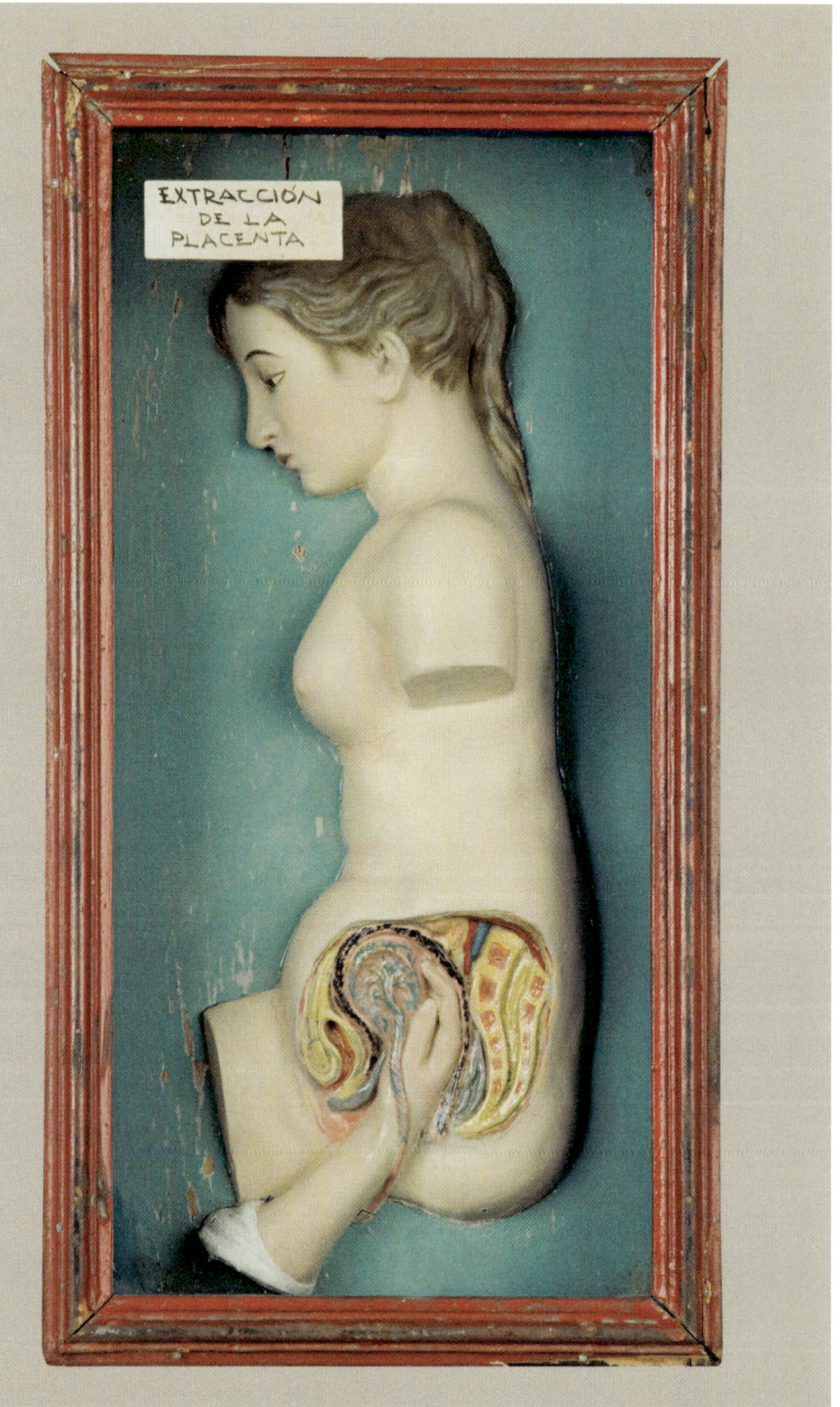

Extraction of the placenta, a plaster relief model from a series illustrating the stages of childbirth (c. 1900) from Barcelona's Museo Roca.

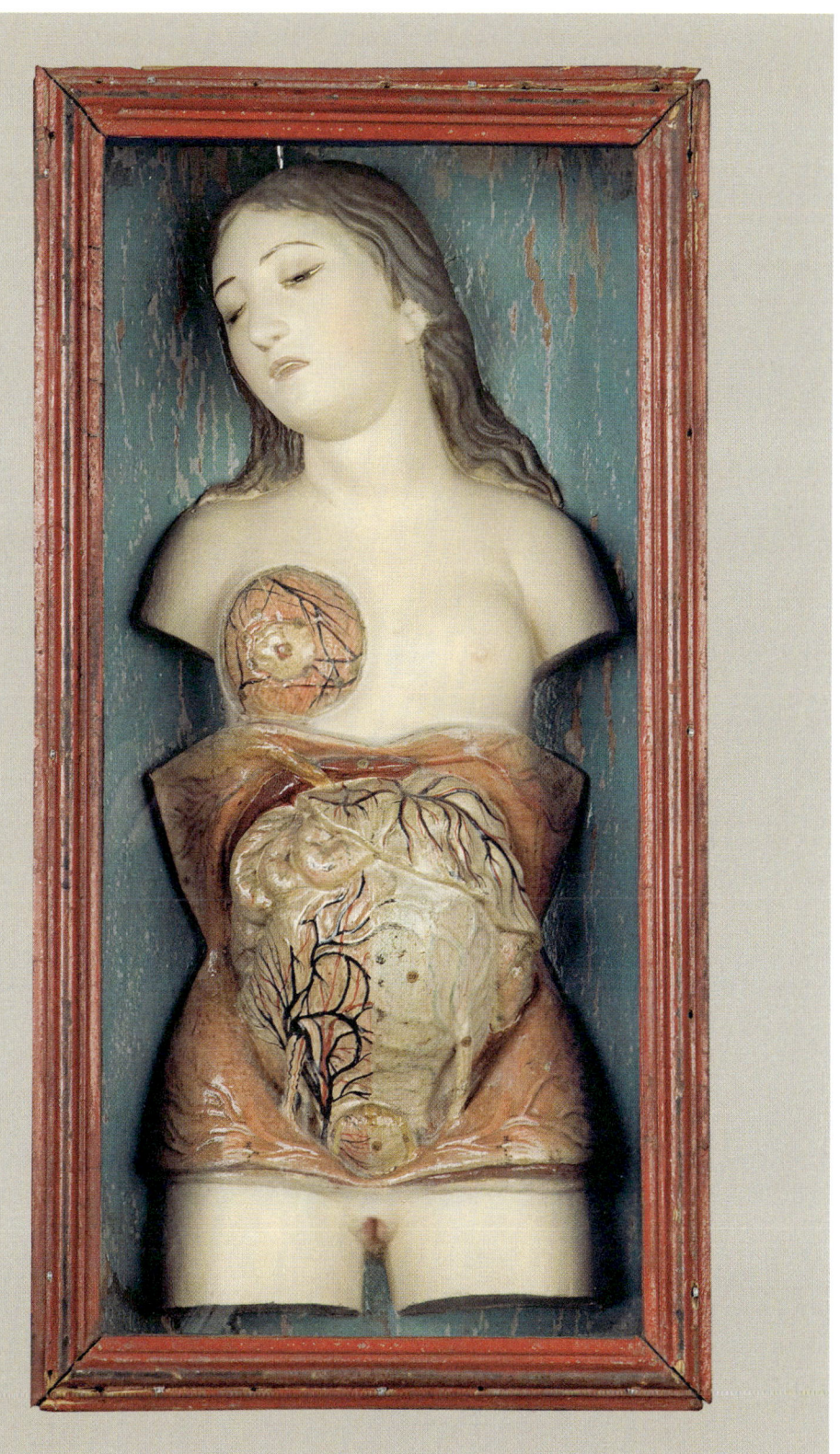

Section of a pregnant woman, plaster relief model from a series illustrating the stages of childbirth (c. 1900), from Barcelona's Museo Roca.

Although these displays can easily be dismissed or demonized as pure voyeurism, they were instructive as well as sensational and titillating. As Maritha Rene Burmeister points out, popular medical museums enjoyed a good reputation for many years, even being recommended in reviews by respected medical journals such as the *Lancet* and the *Medical Times and Gazette*. When Signor Sarti's exhibition—a collection of anatomical waxworks featuring an Anatomical Venus and an Anatomical Adonis—opened in London in 1839, the distinguished literary magazine *Athenaeum* recommended it to 'younger male readers' who wanted to obtain 'a few general ideas on the subject of anatomy, which they may do without labour or disgust'. The study of his models, claimed Sarti, would give the visitor 'the power to communicate intelligibly with his medical advisor' and 'teach him the absolute necessity of putting implicit faith in those men who have made Anatomy and Physiology the study of their lives'. The English surgeon Sir Erasmus Wilson (1809–84) even wrote in 1847 that the exhibition offered 'an unanswerable argument against Atheism'.

fig. 69

fig. 69 A crowd gathers in the Kopstadtplatz, Essen, Germany, in 1889, to see Theatre Rob Melich, a travelling show that presented open-air expositions and a variety of theatrical performances at public fairgrounds. Melich eventually turned his establishment into a cinema.

Such museums became increasingly associated with quack medics, who would sometimes use the horrifying effects of their exhibits as bait for business, and consequently developed a bad reputation. Popular medical museums would often house the more graphic displays in a separate room for men only. A 'doctor' would be available on the premises (for an additional fee, of course) to provide counsel and prescribe patent medications, usually with a mercury base, to gentlemen worried that their own 'night with Venus' might, as the popular maxim warned, have led to 'a lifetime with Mercury'.

By 1857, the *Lancet*, who had once lauded such displays, now described Kahn's Anatomical Museum—one of the largest and longest-running in London—as a place seeking to 'excite the prurient imagination of the young by a stealthy exhibition of obscene prints, figures and books'. Here, 'the ingenious youth, the desponding hypochondriac, or the exhausted roué…his mind duly excited by gazing upon the waxy charms of a "magnificent full-length model of a VENUS, from one of the most eminent of the ancient masters"' would move on to a room filled with models depicting sexual pathology, constituting a kind of 'inner chamber of horrors', where he would stare 'in terrified interest upon a collection of models most revolting, filthy, and disgusting'. When he left that room, he would be given a pamphlet 'which he may study at his leisure, and by the aid of which he may become gradually impressed with the conviction that he is prey to the most terrible and destructive disease, "the results of which are horrible beyond description", from which there is no prospect of safety but in the mysterious science of that extraordinary man.'

One of the best known of all popular medical museums was opened in Paris in 1856 by the French self-styled doctor Pierre Spitzner under the banner 'Art, Science, Progress!' It remained in its original Paris location until the 1880s when, after a devastating fire, it went on the road to travel the great fairgrounds of France, Britain, Germany, Holland, and Belgium, finally closing its doors during World War II. The 1908 poster for Spitzner's *Grand Muséum d'Anatomie* announced:

fig. 70 The interior of Museo Roca, a popular anatomical museum in Barcelona's red-light district. Señor Roca was a successful fairground entrepreneur in the early twentieth century.

fig. 71 A still from the 2012 silent Spanish film Blancanieves (Snow White). Paying customers queue for the chance to kiss and awaken the sleeping beauty on display at a fair in a glass coffin.

fig. 70

fig. 71

> *Ladies and Gentlemen: from birth until death the road is long. A creature's arrival on the scene of life and its development before reaching its definitive form represent astonishing problems. Anatomical science consists of this: Seeking the secret of life in death! In the injured organ, seeking the cause of the disease in order to alleviate suffering. That is what a scientist does! For you the People, it's another matter. For you, anatomy is a reality about which your mind requires knowledge. Anatomical waxworks will teach you to understand yourself physically. Thus you will be able to contemplate your strengths and your weaknesses. Pathology will produce in you salutary dread.*

Spitzner toured with wax moulages, models intended to demonstrate the ravages of alcoholism and venereal disease, particularly syphilis, which was raging across Europe. The collection is unique in that it remained a viable business for so long, and in that it still exists—now at the University of Montpellier,

THE MOST INCREDIBLE EXAMPLE OF SADIST-SURREALIST FANTASY IS TO BE FOUND AMONG THE REPRESENTATIONS OF THE VARIOUS PHRASES OF CHILDBIRTH AND GYNAECOLOGICAL OPERATIONS. A COMPLETE MODEL OF A PATIENT UNDERGOING A CAESAREAN SECTION LIES WITH HER EYES WIDE OPEN, HER FACE DISTORTED BY PAIN, HER HAIR IMPECCABLE, HER CALVES TIED TOGETHER, DRESSED IN A LONG, LACE NIGHTGOWN, WHICH IS OPEN ONLY AT THE PART OF HER BODY WHICH HAS BEEN CUT OPEN BY A SCALPEL, WHERE THE BABY APPEARS. FOUR MALE HANDS ARE PLACED ON HER BODY (TWO OPERATING, TWO HOLDING HER WAIST): FINE WAX HANDS WITH MANICURED NAILS, GHOSTLY HANDS SINCE THEY ARE NOT SUPPORTED BY ARMS BUT ADORNED ONLY WITH WHITE CUFFS AND WITH THE ENDS OF THE SLEEVES OF A BLACK JACKET, AS THOUGH THE WHOLE CEREMONY WAS BEING HELD BY PEOPLE IN EVENING DRESS.

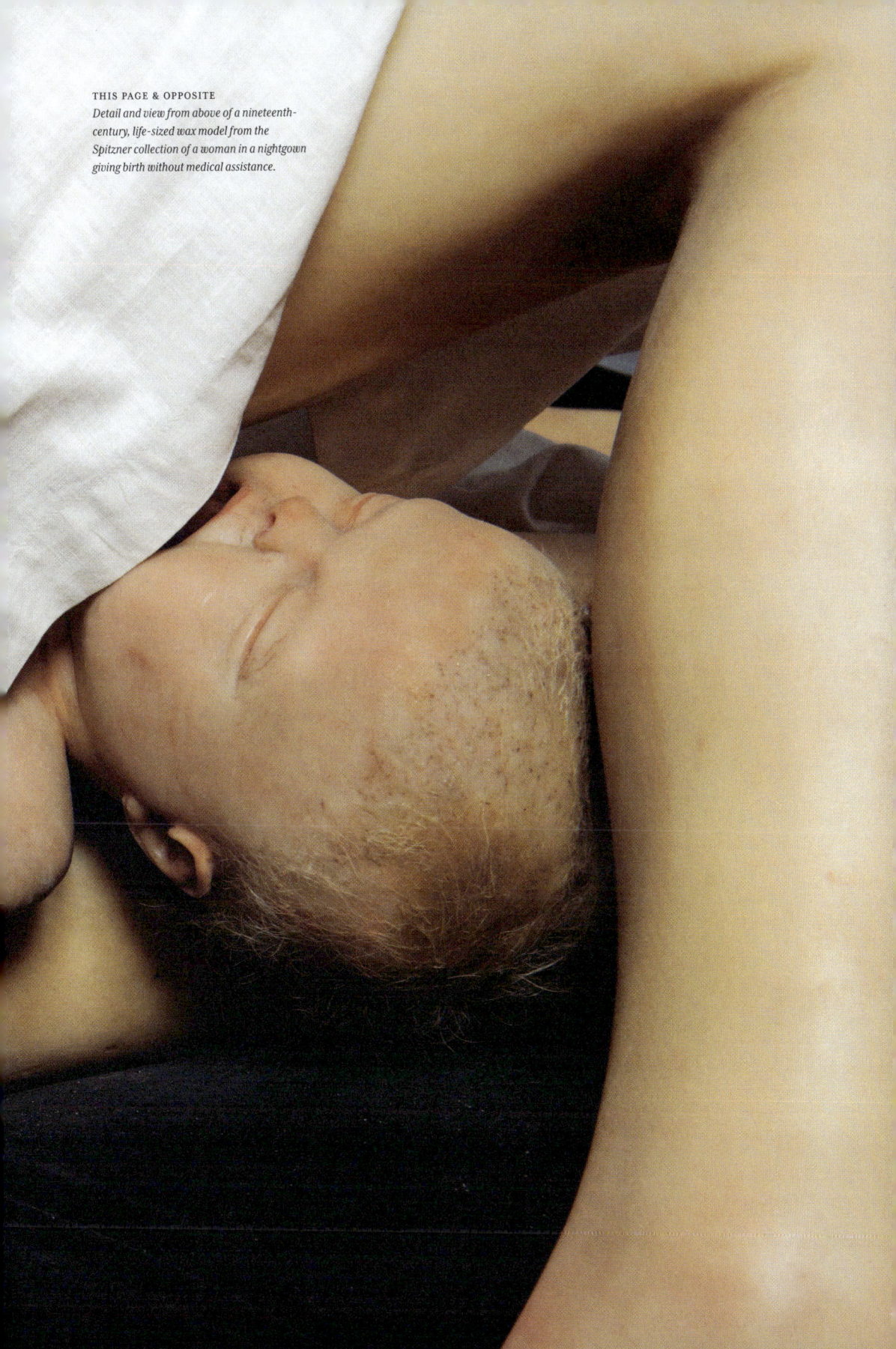

THIS PAGE & OPPOSITE
Detail and view from above of a nineteenth-century, life-sized wax model from the Spitzner collection of a woman in a nightgown giving birth without medical assistance.

THIS PAGE & OPPOSITE *Life-sized wax model of a well-coiffed woman enduring a difficult assisted birth conducted by phantom hands. From the Spitzner collection.*

France—in a relatively intact state and with its remarkable pieces in excellent repair, having undergone an expert restoration.

Spitzner's collection exhibited a variety of full-scale, recumbent female figures, including an Anatomical Venus dissectible into forty pieces and a caesarean section performed by phantom hands on a bound woman. The most spectacular and memorable of these figures was the beautiful *Sleeping Venus* (see page 168) that evokes Philippe Curtius's *Sleeping Beauty* (see page 175), an automatized waxwork whose breast rose and fell subtly, as if she were alive and breathing. This breathing Venus has been described by Kathryn A. Hoffmann as just one of the 'various imaginaries of death and somnolence that developed

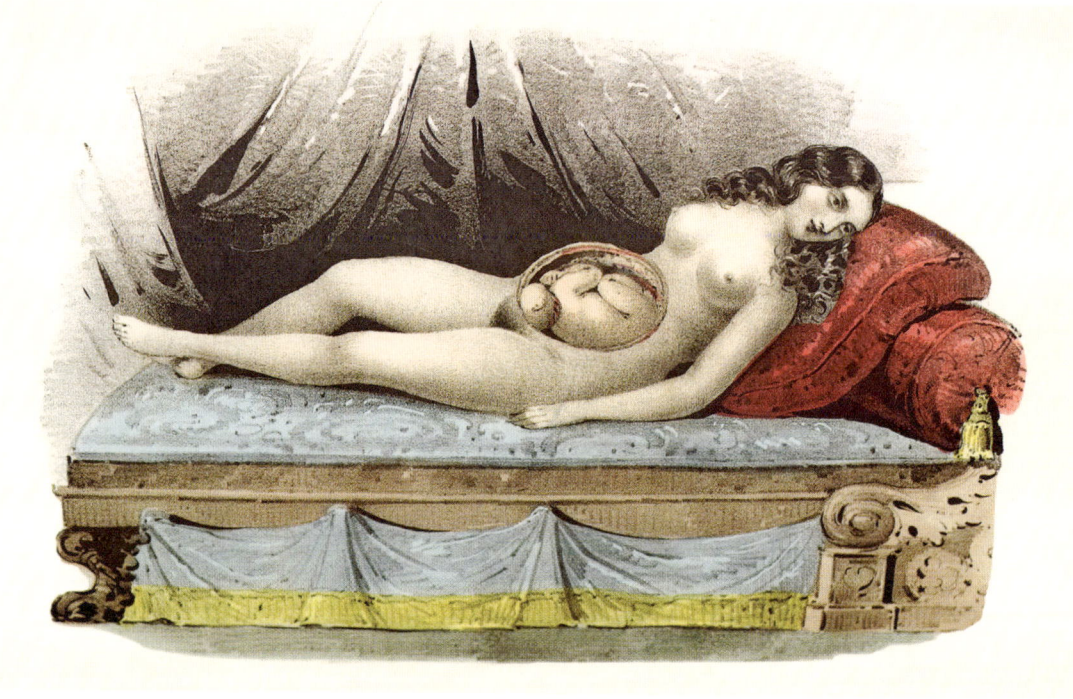

fig. 72

fig. 72 Illustration from Beach's An Improved System of Midwifery *(1851), depicting a pregnant woman with a cutaway belly reclining on a chaise.*

across the decades and formed part of fairground entertainments'. Indeed, as she explains, women in glass boxes were frequent sights at the fairground, and part of the pleasure of these displays would be the game of trying to determine whether the girl was real, waxwork, or automated waxwork. These girls in boxes might be sleeping beauty *tableaux vivants*, which continued to be displayed at least into the 1960s, or 'a mesmerized girl about to reveal her trip in the ether to far-away lands, a somnambulist about to walk off her couch with the next breath, a sick sensitive, or a clairvoyant in a coma'. In the black-and-white Spanish silent film *Blancanieves* (2012), she is a virginal Snow White displayed in a glitteringly beautiful glass coffin, with a line of onlookers queuing to pay for the chance to see whether their kiss will be the one that wakes her (see page 149).

Sleeping beauties were a popular theme in the vogue for posing and staging the dead as if beautifully asleep in nineteenth-century post-mortem photography. French stage and film actress Sarah Bernhardt (1844–1923) had herself photographed as a sleeping beauty in 1864, and again in 1880, in a coffin she had owned—and reported that she had also often slept in to help her to comprehend and realize her many tragic roles—since her somewhat sickly childhood. Arguably the most celebrated actress of her day, she was especially beloved for her romantic death scenes in works such as *La Dame aux Camélias*, *Cleopatra*, and *Hamlet*. According to Ockman and Silver in *Sarah Bernhardt: The Art of High Drama* (2005), Bernhardt was the queen of morbid sexuality:

fig. 73

> By the early 1880s, [Sarah Bernhardt's] talent for dying was so remarked on that a final death agony was practically mandatory. Rhapsodic critics insisted that she never died the same way twice. For the next forty years, she died nightly, and sometimes twice a day...

Real human corpses were a popular entertainment at the Paris Morgue, a major tourist attraction from the nineteenth century until it closed in 1907 'out of concern for public morality'. Bodies were laid out on marble slabs, their genitals covered and clothes hanging up beside them, ostensibly for the purpose of identification. The morgue would attract as many as 40,000 people a day when bodies reported on by newspapers were displayed, with the corpses of women

fig. 73 Early-twentieth-century postcard showing a sleeping beauty in a glass box as part of a fairground display. To the right of the sleeping beauty is a tableau probably based on Fuseli's 1781 painting The Nightmare (see fig. 56, page 130).

MY BEDROOM WAS QUITE TINY. THE BIG BAMBOO BED TOOK UP ALL THE ROOM. IN FRONT OF THE WINDOW WAS MY COFFIN, WHERE I FREQUENTLY INSTALLED MYSELF TO STUDY MY PARTS. THEREFORE, WHEN I TOOK MY SISTER TO MY HOME I FOUND IT QUITE NATURAL TO SLEEP EVERY NIGHT IN THIS LITTLE BED OF WHITE SATIN WHICH WAS TO BE MY LAST COUCH, AND TO PUT MY SISTER IN THE BIG BAMBOO BED, UNDER THE LACE HANGINGS.

Excerpt from French actress Sarah Bernhardt's autobiography My Double Life: The Memoirs of Sarah Bernhardt *(1907).*

[3] Achille Melandri's albumen print cabinet card (c. 1880) of Sarah Bernhardt posing as if dead in a coffin she had owned, and reported that she had often slept in, since her sickly childhood.

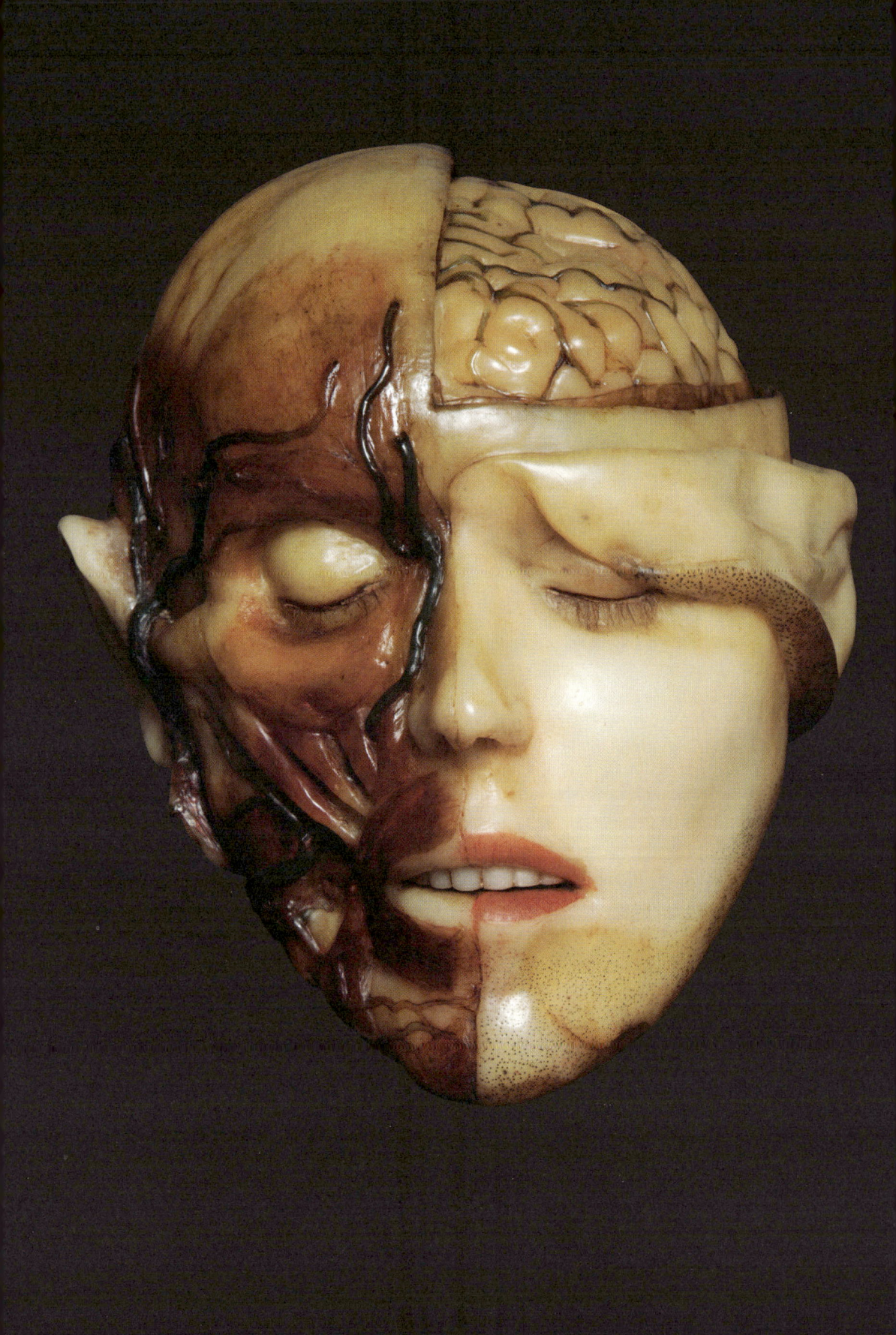

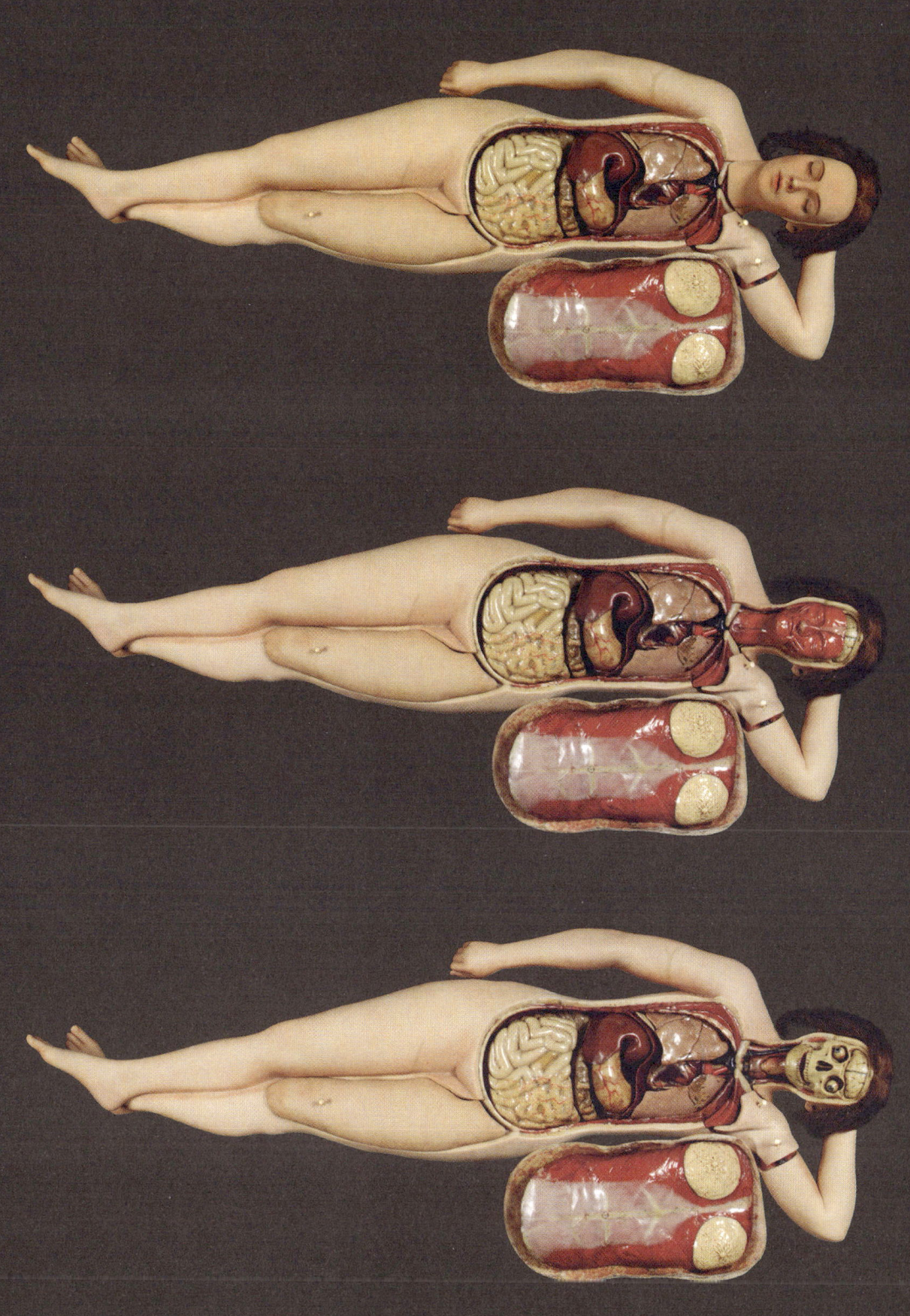

THIS PAGE & OPPOSITE *The face of the Spitzner collection's life-sized forty-piece Anatomical Venus (see page 161), shown in its intact form and dissected state.*

In the middle of the entrance to the museum was a woman who was the cashier, then on one side there was a man's skeleton and the skeleton of a monkey, and on the other side there was a representation of Siamese twins. And in the interior one saw a rather dramatic and terrifying series of anatomical casts in wax which represented the dramas and horrors of syphilis, the dramas, deformations. And all this in the midst of the artificial gaiety of the fair. The contrast was so striking that it made a powerful impression on me. All the 'Sleeping Venuses' that I have made, come from there. Even the one in London, at the Tate Gallery. It is an exact copy of the Sleeping Venus in the Spitzner Museum, but with Greek temples or dressmaker's dummies, and the like. It is different, certainly, but the underlying feeling is the same…

In a 1973 interview with Renilde Hammacher, surrealist painter Paul Delvaux (1897–1994) describes his impressions upon viewing Spitzner's Sleeping Venus at the Brussels Fair, Belgium, around 1930.

Sleeping Venus (La Vénus endormie) by Paul Delvaux (1944). Several of Delvaux's paintings were inspired by the Spitzner collection's Sleeping Venus.

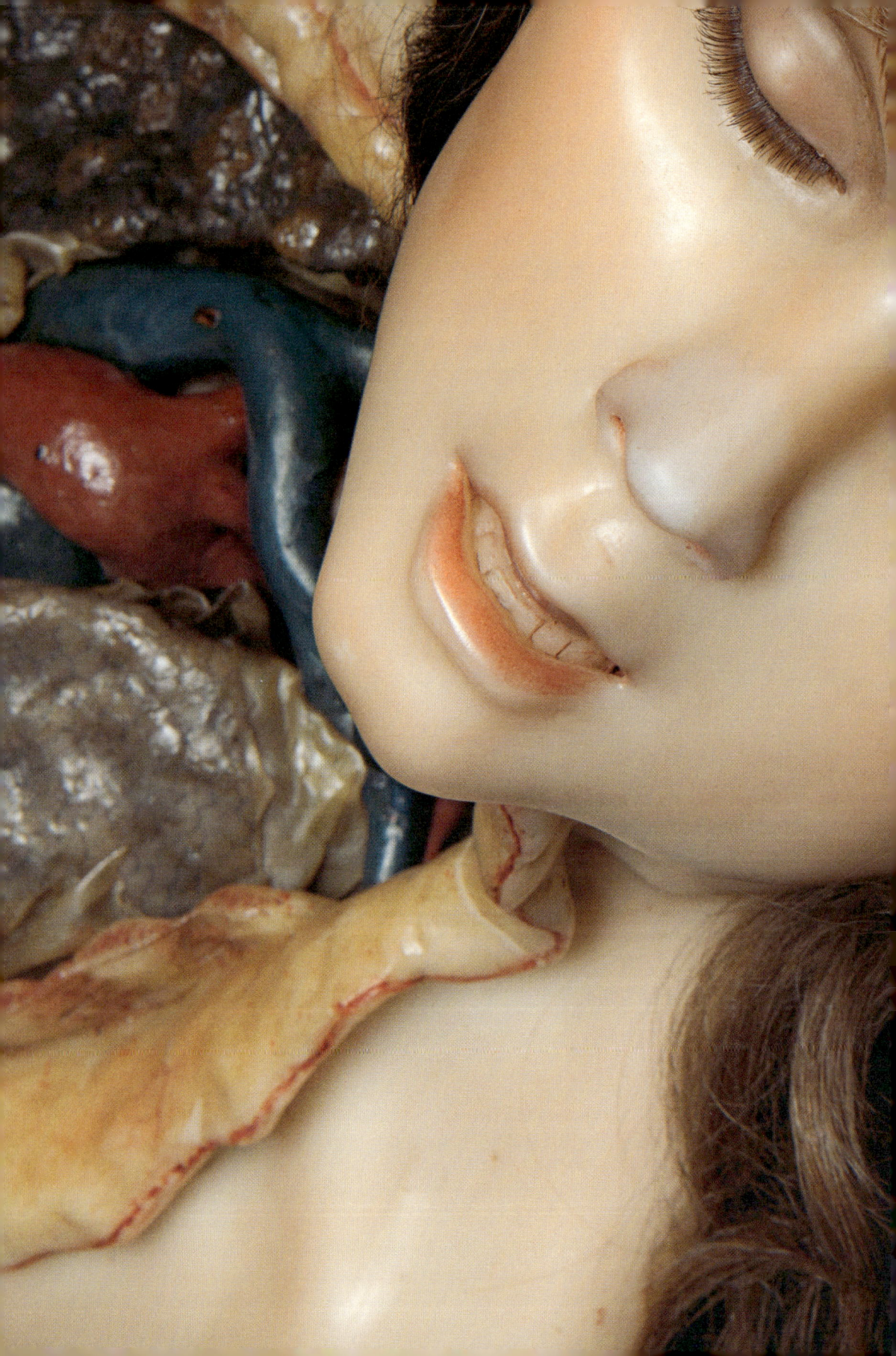

PREVIOUS *Detail of a late-nineteenth-century, life-sized wax bust of a dissected woman with tuberculosis. Part of the Spitzner collection.*

[3] The Spitzner collection's Sleeping Venus. Spitzner claimed in 1874 to have won a medal for the mechanized waxwork; the breast subtly rises and falls as if the model were breathing.

and children drawing the biggest crowds. The morgue was listed as a must-see 'Museum of the Real' in guidebooks of the era. The French daily newspaper *La Presse* reported in 1907:

> The Morgue is considered in Paris like a museum that is much more fascinating than even a wax museum because the people displayed are real flesh and blood.

It was even possible to check out body parts for private use, as Théodore Géricault (1791–1824) did for studies for *The Raft of the Medusa* (1819). The morgue also featured in popular novels, such as George du Maurier's *Trilby* (1894). A passage from Émile Zola's novel *Thérèse Raquin* (1867) captures latent eroticism in a description of the murderer Laurent's regular visits to the Paris Morgue, where he attempts to find his victim's body:

fig. 74 Plaster death mask of L'Inconnue de la Seine (c. 1880).

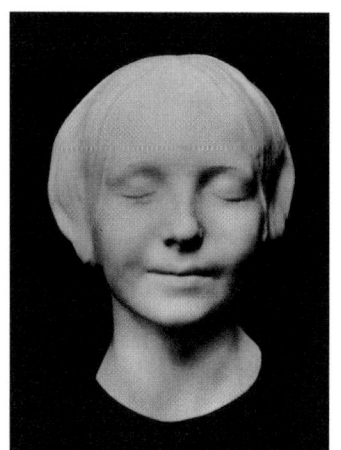
fig. 74

fig. 75

fig. 76

fig. 75 A student drawing the death mask of L'Inconnue de la Seine (c. 1890).

fig. 76 Asmund Laerdal practising CPR on Rerucci Anne, possibly the most kissed face of all time.

> When there were no drowned persons on the back row of slabs, he breathed at ease; his repugnance was not so great. He then became a simple spectator, who took strange pleasure in looking death by violence in the face, in its lugubriously fantastic and grotesque attitudes. This sight amused him, particularly when there were women there displaying their bare bosoms. These nudities, brutally exposed, bloodstained, and in places bored with holes, attracted and detained him.
>
> Once he saw a young woman of twenty there, a child of the people, broad and strong, who seemed asleep on the stone. Her fresh, plump, white form displayed the most delicate softness of tint. She was half smiling, with her head slightly inclined on one side. Around her neck she had a black band, which gave her a sort of necklet of shadow. She was a girl who had hanged herself in a fit of love madness.

One of the most enduring stories relating to the Paris Morgue is that of *L'Inconnue de la Seine*, or 'the unknown woman of the Seine'. In this story—which may be apocryphal—a beautiful young suicide of about sixteen years of age was pulled from the Seine. Her body was displayed at the Paris Morgue, where a pathologist, captivated by her enigmatic 'Mona Lisa smile', made a death mask of her face. A generation of bohemians hung plaster replicas of *L'Inconnue*'s death mask in their homes and some even argued that her face served as 'the erotic ideal' of her period.

L'Inconnue de la Seine's afterlife extends to the present; her serene visage was used as the model for the face of 'Resusci Anne', the doll used to teach CPR (cardiopulmonary resuscitation). After saving his young son from drowning in 1955, Norwegian toy manufacturer Asmund Laerdal sought to develop a realistic rubber model with which to simulate the new life-saving method. He decided that the calm and unthreatening appearance of the ubiquitously reproduced plaster mask of *L'Inconnue* would make a perfect model for students of first-aid. Launched in 1960, Resusci Anne has been called 'the most kissed face of all time'.

fig. 77 Oil painting of body parts borrowed from the Paris morgue, used as studies by Théodore Géricault when painting The Raft of the Medusa *(1819).*

fig. 77

fig. 78

Although the British Obscene Publications Act of 1857 and the American Comstock Laws of 1873 closed many popular anatomical museums by the end of the nineteenth century, feminine death, illness, and inertia were kept in fashion by the Romantic arts, especially the Pre-Raphaelites; by the cult of mourning begun by Queen Victoria upon the death of her husband Prince Albert; and by the melodramatic Gothic novels of the day. The persistence of this allure is apparent in the corpse of transgressively sexual Laura Palmer—the central subject of David Lynch's cult television series *Twin Peaks* (1990–91)—and in artist Cornelia Parker's installation *The Maybe* (1995), in which Tilda Swinton slept in an elevated glass box for seven days. Eerie waxworks are still staples in the American horror movie, and come alive in Ray Bradbury's seminal novel *Something Wicked This Way Comes* (1962). Gunther Von Hagens's international 'Body Worlds' exhibitions—with their somewhat kitschily posed, real 'plastinated' corpses and anatomical preparations—embody the spirit of popular early anatomy shows; clearly, the fascination with human mortality which was such an intrinsic part of the Anatomical Venus's creation still endures to this day.

fig. 78 Ophelia *(1851–52), by Sir John Everett Millais. The model for Ophelia, Elizabeth Siddal, posed in a bath full of tepid water over a four-month period, becoming severely ill as a result.*

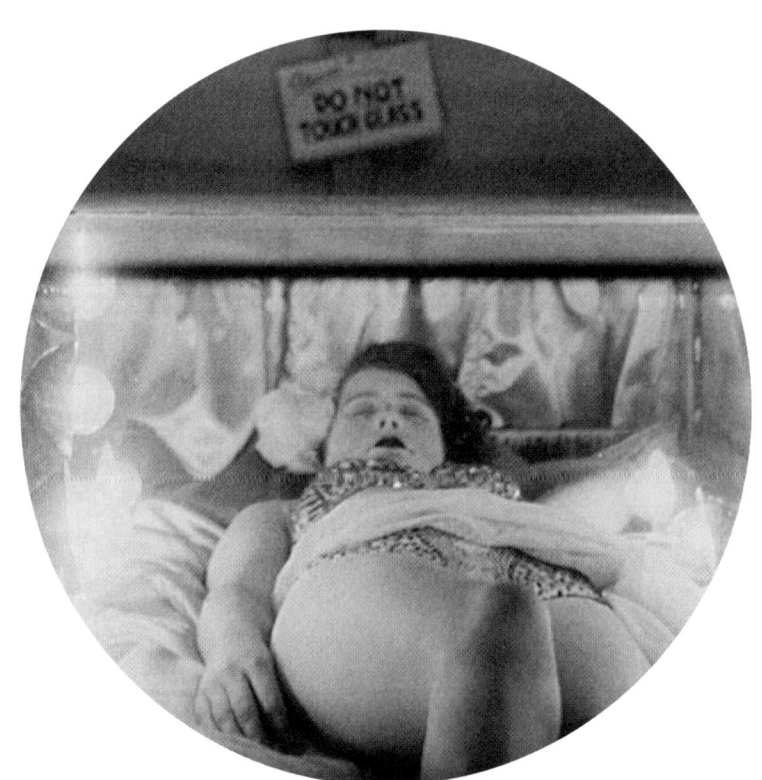

TOP Façade of 7 Sleeping Beauty show at Cannon Hill Park's Tulip Festival, Birmingham, England (1962). BOTTOM Price's Sleeping Beauty: a live woman posing in a glass box (1950s).

AMONG THE SHOW TYPES THAT MIGHT USE A RECUMBENT GIRL IN A COFFIN WERE LIVING IMITATIONS OF MEMORIAL PHOTOGRAPHY.... SOME DISPLAYED DEAD YOUNG GIRLS AS SLEEPING BEAUTIES, SWATHED IN YARDS OF WHITE SATIN, SILK, AND TULLE, FLOWERS CLASPED IN DEAD HANDS OR STREWN THROUGHOUT THE CASKETS, PALE FACES FRAMED BY DARK HAIR.

THE·DEATH,
THEN,·OF·A
BEAUTIFUL
WOMAN·IS,
UNQUESTION-
ABLY,
THE·MOST
POETICAL
TOPIC·IN
THE·WORLD.

An excerpt from Edgar Allan Poe's 1846 essay 'The Philosophy of Composition', published in Graham's Magazine Vol XXVIII. Poe lost his wife to consumption when she was only twenty-four.

The Sleeping Beauty, a breathing wax model by Swiss physician and master wax sculptor Philippe Curtius. The model pictured here is a 1925 replica cast from the same mould, after the original wax piece from 1767 was destroyed in a fire. British writer Marina Warner says of this piece: 'The illusion of permanent sleep is invoked to deny the reality of death… The Sleeping Beauty functions as anti-memento mori…. she promises immortality as the suspension of time.' (Phantasmagoria, 2006)

OVERLEAF *Dissectible life-sized wax Anatomical Venus, created around 1930 by the workshop of Rudolf Pohl. The model was exhibited at a fairground museum as part of Oktoberfest in both 1933 and 1934.*

CHAPTER FOUR

ECSTASY, FETISHISM, AND·DOLL WORSHIP

(179)

PREVIOUS
Detail of Gian Lorenzo Bernini's marble sculpture The Rape of Proserpina *(1621–22), depicting the kidnapping of Proserpina (in Greek, Persephone) by Pluto, who takes her to Hades.*

fig. 79 *An Anatomical Venus from La Specola idly plays with her long golden braid made of human hair.*

The Medici Venus looks perplexing and sometimes even troubling to the contemporary eye, a response that is largely due to the startling juxtaposition between the abjectness of her innards and the erotic charge of her hyper-beautiful exterior. This odd sense is heightened by the inclusion of details that, to today's sensibility, seem superfluous to her educational aims: her cascade of lustrous hair, life-like glass eyes, pearl necklace (a memento mori symbol that also covers the seam that attaches her head to her torso), and bed of luxurious silk and satin. Such artful and seductive details seem to undermine her scientific integrity, but it is the intertwining of anatomy, art, and religion that unlocks the meaning of her provocative appeal, and that of other sexually arousing models and dolls created or mythologized by men.

The Medici Venus's enigmatic expression and swooning posture are suggestive of ecstasy, which today is understood as sexual. We can infer that at the time of her creation she was not regarded as particularly indecent from the fact that, upon opening to the general public, including women and children,

fig. 79

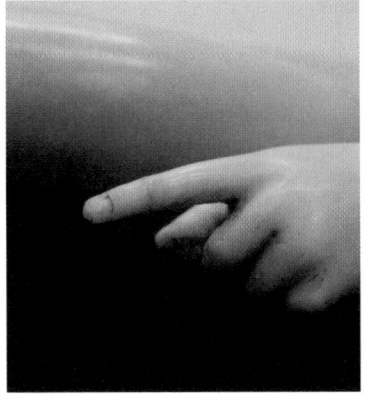

fig. 80

fig. 81

fig. 80 *The hand of a half-sized Anatomical Venus with index finger extended in a life-like gesture.*

fig. 81 *The* Venerina *wears a pearl necklace wrapped tightly around her neck. It served three functions: aesthetic, symbolic (of the passing vanity of human life), and concealing (of the seam that attaches the head to the torso).*

the wax models were the most popular exhibits, eliciting neither criticism nor public outcry. There are scores of comments by visitors to the museum who spoke highly of the collection, and we know that copies were commissioned by a variety of other museums. Clearly, something has changed since the time of Venus's creation in such a way as to render her strange and sexually charged.

It is likely that a different understanding of the ecstatic than our own influenced Venus's reception. The ecstatic was understood at that time not merely as a profane, sensual experience, but as an expression of the sacred: a mystical experience. Depictions of saints and martyrs in attitudes of ecstatic release fill the churches of Italy and other Catholic countries. Probably the best known of these is Gian Lorenzo Bernini's life-sized white marble masterpiece, the *Ecstasy of Saint Teresa* (1647–52) at Santa Maria della Vittoria in Rome. By investigating the plurality of responses people of other times have shown to this seminal work, we can appreciate how reactions to such imagery have changed over time.

Bernini (1598–1680)—who was, by all accounts, a devout Catholic—used a specific passage from St Teresa's sixteenth-century autobiography, *The Life of*

Saint Teresa of Avila, as the departure point for his statue; in it, the saint explains that, while singing the hymn *Veni, Creator Spiritus* (Come, Creator Spirit), she saw a long spear of gold with an iron tip in the hand of an extraordinarily beautiful angel who had appeared to her:

> *Very close to me…an angel appeared in human form…he was not tall…but very beautiful and his face was so aflame that he appeared like one of those superior angels who look as though they are completely on fire.… In his hands I saw a large golden spear and at its iron tip there seemed to be a point of fire. He appeared to me to be thrusting it at times into my heart, and to pierce my very entrails; when he drew it out, he seemed to draw them out also, and to leave me all on fire with a great love of God. The pain was so great, that it made me moan; and yet so surpassing was the sweetness of this excessive pain, that I could not wish to be rid of it.… The pain is not bodily, but spiritual; though the body has its share in it, even a large one. It is a caressing of love so sweet which now takes place between the soul and God.…*

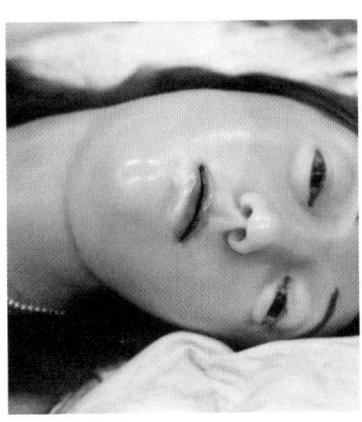

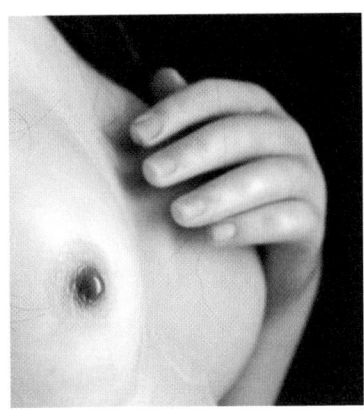

fig. 82

fig. 83

fig. 84

This passage describes Saint Teresa's 'transverberation', or piercing through the heart. The account reads as extremely suggestive to those outside the faith, but there is no indication that the young Teresa's diary is considered impure: after all, she was elevated to sainthood, and her diary is routinely read by the devout. Bernini's sculpture was placed in the church without censure and celebrated as a religious masterpiece. The diary entry is even displayed at the church in both English and Italian, and across the chapel there is a life-sized wax effigy of Saint Victoria in a glass case, demonstrating a similar attitude of ambiguous ecstasy.

We also know that by the eighteenth century expressions of religious rapture sometimes invited more familiar, corporeal interpretations, as in the sly comment by French scholar and magistrate Charles de Brosses (1709–77) upon seeing Bernini's statue of Saint Teresa in 1739: 'If this is divine love, I know all about it.'

How are we to understand such ecstatic representations today? Perhaps the heart of the confusion lies in the notion of the ecstatic as either religious or

fig. 82 The expression on the face of the Medici Venus is simultaneously relaxed and ecstatic.

fig. 83 A wax model of a woman in the ninth month of pregnancy (c. 1880) from the workshop of Gustav Zeiller. She does not exhibit the enlarged, darkened areolae characteristic of her condition.

fig. 84 An Anatomical Venus from La Specola, with an arched neck and a braid of human hair draped over her shoulder.

HE APPEARED TO ME TO BE THRUSTING IT AT TIMES INTO MY HEART, AND TO PIERCE MY VERY ENTRAILS; WHEN HE DREW IT OUT, HE SEEMED TO DRAW THEM OUT ALSO, AND TO LEAVE ME ALL ON FIRE WITH A GREAT LOVE OF GOD. THE PAIN WAS SO GREAT, THAT IT MADE ME MOAN; AND YET SO SURPASSING WAS THE SWEETNESS OF THIS EXCESSIVE PAIN, THAT I COULD NOT WISH TO BE RID OF IT... THE PAIN IS NOT BODILY, BUT SPIRITUAL; THOUGH THE BODY HAS ITS SHARE IN IT, EVEN A LARGE ONE. IT IS A CARESSING OF LOVE SO SWEET WHICH NOW TAKES PLACE BETWEEN THE SOUL AND GOD.

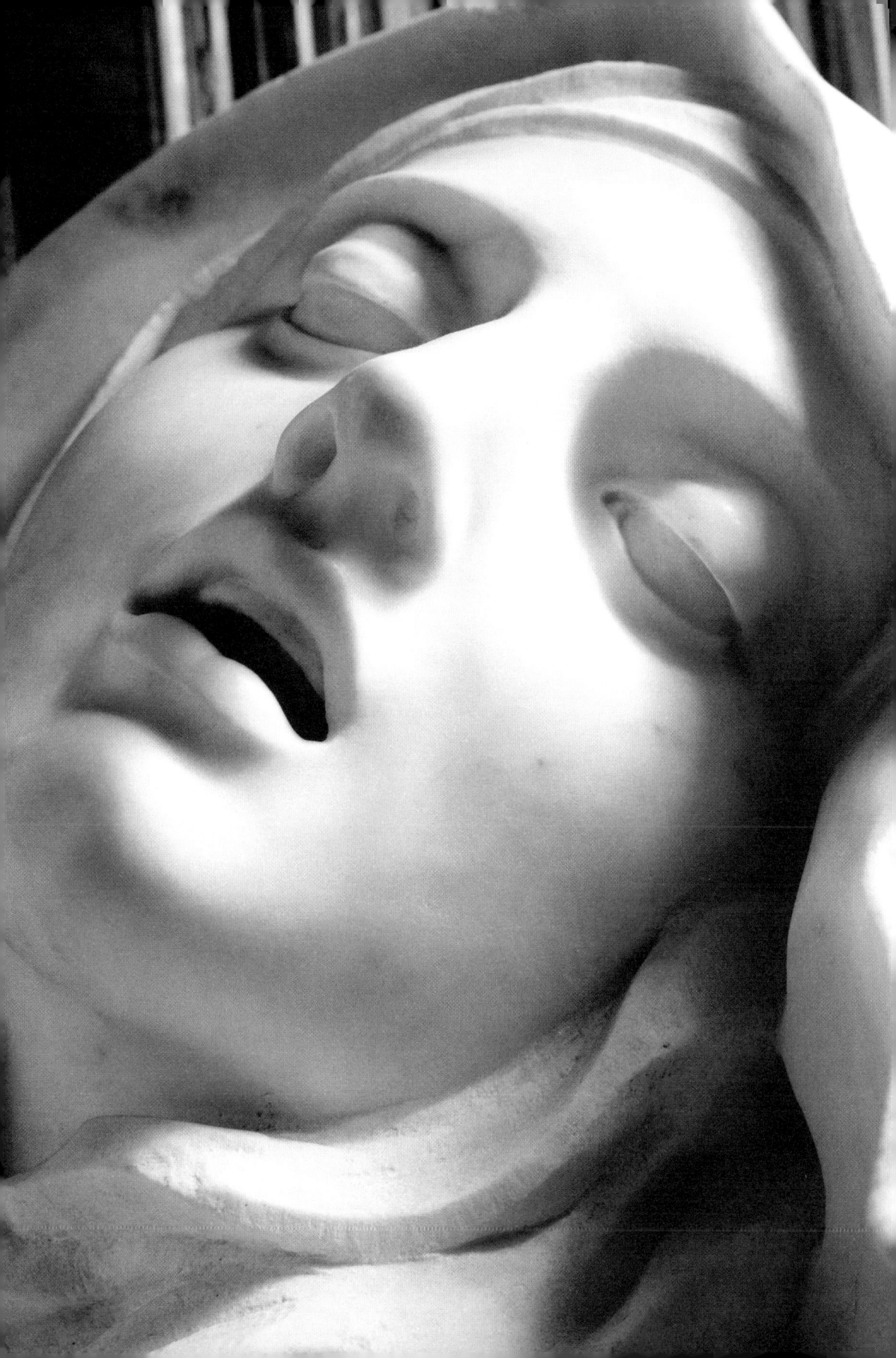

sexual, sacred or profane, where once it was understood to be both. In *Eroticism* (1957), the French intellectual Georges Bataille (1897–1962) argues that sexuality itself was a part of religious expression until Christianity banished it from that domain. The sexual act—and other forms of ecstatic release, such as intoxication or ritual chanting and dancing—serves to dissolve the boundary of the individual; we lose ourselves in a larger 'other', engendering a sense of becoming one with the universe, or losing our sense of individuality. The very etymology of the word ecstasy lies in the Greek *ek*, meaning 'out of', and *stasis* meaning 'standing'. This escape from isolated consciousness is experienced as a sort of transcendent bliss beyond the powers of language or description. It is, in Bataille's words, 'divine life sought through death of the self' that is at the heart of the ecstatic

fig. 85

fig. 85 Bernini's funerary monument of the pious Blessed Ludovica Albertoni (1674), shown at the moment of a mystical communion with God. From the Altieri Chapel, Church of San Francesco at Ripa, Rome, Italy.

experience. It is our way of returning to the state of grace that existed before we were banished from the Garden of Eden; before we were divided from the universe by our self-awareness, our language and propensity for abstraction, our sense of shame, and the foreknowledge of our own death—arguably the very thing that separates us from other animals. This is the enlightened epiphany described by mystics of all religious backgrounds. It is no wonder that sexuality retains a touch of the numinous (that is, the spiritual or supernatural), if one rendered strange by our attempts to focus it squarely in the world of the body and the senses, and to separate it from more mysterious domains.

It is this completely blissful state of religious transcendence, physical as well as spiritual, that Bernini so effectively reproduced in his statue. In being pierced by an angel's spear, Saint Teresa's love of God is consummated, and she enters a state of spiritual marriage. Her new knowledge of the divine arrives with a 'little death' that depicts the same intense transport into a non-material realm as sexual orgasm is understood to do today. Bernini encapsulates the beautiful, ephemeral moment in which the saint is caught between heaven and earth, matter and spirit.

What L. J. Jordanova describes in *Sexual Visions* (1989) as an 'ambiguous mixture of sexual and religious ecstasy' in Bernini's sculpture says more about our attitudes today than attitudes at the time of the Anatomical Venus's creation. It is a historically new stance that makes strictly private—and often deviant or even demonic—the ways in which mankind has formerly found ecstatic union with the divine: sexuality, intoxication, and rituals encouraging loss of ego awareness. Transcendence comes by allowing our personal boundaries to be

fig. 86 A postcard featuring a still from the first episode of the silent film serial Who is Number One? (1917), starring Kathleen Clifford as Aimee Villon.

fig. 86

pierced or blurred in ecstatic release; for Saint Teresa, standing outside herself creates the space for God to penetrate her completely. In the case of the Medici Venus, her ecstatic attitude invites men of science to penetrate the secrets of Nature, and thus to take the place of the divine creator. As Fulvio Simoni notes in his article *'Anatomie conturbanti'* ('Perturbing Anatomies'):

> ...scrutinizing the most secret parts of the body meant becoming aware of the complexity of the structures and mechanisms that made it work; but it also meant contemplating God, the Great Watchmaker of the mechanists and deists, through the most marvellous product of his creation.

Paradoxically, excluding the non-rational and the numinous from serious contemplation or discussion gives both those states power. Banished from

TOP *Anatomy of the heart by Enrique Simonet y Lombardo (1890).* BOTTOM *The Anatomist by Gabriel von Max (1869).*

These holy virgins are dying of love, and the little death of sexual pleasure is confounded with the final death of the body... The dead body becomes in its turn an object of desire.

Excerpt from The Hour of Our Death (1981) by French historian Philippe Ariès. The book traces changing Western attitudes to death over the last one thousand years.

figs 87–92 Illustrations by Charles Raymond from Venus im Pelz (1870) by Leopold von Sacher-Masoch, published in English in 1928 as Venus in Furs. The book makes clear the numinosity that activates the sexual fetish: 'I saw sensuality as sacred, indeed the only sacredness, I saw woman and her beauty as divine since her calling is the most important task of existence: the propagation of the species.'

being publicly condoned and experienced, a sense of communion with something beyond material nature can attach itself to transgressive acts or external objects, which come to occupy the magical position between matter and spirit. One manifestation of this is the idea of the fetish object and the sexual fetish.

A fetish is a man-made object believed to have supernatural powers. The word was originally coined to describe cult objects from West Africa, but can also be applied to powerful objects of Catholicism, such as consecrated hosts and ex-votos. The idea of the fetish was expanded by French psychologist Alfred Binet (1857–1911), who used the term 'sexual fetish' in 1887 to describe the phenomenon of sexual attachment to an object or body part not traditionally viewed as sexual. It should come as no surprise that the boundary-dissolving, mysterious realms of love and sexuality continue to provide a home for the numinous in the form of the sexual fetish.

fig. 87

fig. 88

fig. 89

Many sexual fetishes were painstakingly named and described by the Austro-German psychiatrist Richard von Krafft-Ebing (1840–1902) in *Psychopathia Sexualis* (1886). Besides popularizing the terms 'sadism' (from the proclivities of the Marquis de Sade) and 'masochism' (from Leopold von Sacher-Masoch's 1870 book *Venus im Pelz*), he devotes a few paragraphs to a survey of the historical record of the fetishization of feminine statues. Krafft-Ebing begins in ancient times, relating the story of Clisyphus 'who violated the statue of a goddess in the Temple of Samos after having placed a piece of meat in a certain part', and brings the fetish into the modern world with a story from 1877 about 'a gardener who fell in love with a statue of the Venus de Milo and was discovered attempting coitus with her'. This particular fetish is termed 'agalmatophilia' (Greek *agalma*, statue, and *philia*, love) and can be described as a sexual attraction to a statue, doll, mannequin or other figurative object. This can include a desire for actual sexual contact with the object, or a fantasy of having sexual or non-sexual encounters with an animate or inanimate instance of the object.

One sub-category of agalmatophilia is Pygmalionism, or the phenomenon of falling in love with an object of one's own creation. This is based on the Ancient Greek myth in which the sculptor Pygmalion falls in love with a statue he has made of a woman. Aphrodite, Greek counterpart of the Roman Venus, takes mercy on the lovelorn sculptor and grants life to his statue, named 'Galatea' in later retellings. Agalmatophilia and Pygmalionism were popular artistic and literary subjects of the nineteenth century, notably in paintings by Jean-Léon Gérôme (1824–1904) and Edward Burne-Jones (1833–98), and in E. T. A. Hoffmann's story *The Sandman* (1816), in which a young man disappointed in love is driven to madness by his passion for an automaton, Olympia.

fig. 90

fig. 91

fig. 92

There are also some rather astounding real-world cases of agalmatophilia and Pygmalionism, often containing a hint of necrophilia, the sexual desire for dead bodies. At the base of each of these paraphilias is a desire forever to preserve, to defeat death, and to possess and control the female body. Just as the Anatomical Venus masks the dead bodies on which she is modelled, so the dolls created by men who wish more fully to possess the women they desire mask the death or disappointment of the love object. An energy is maintained by ensuring that the object occupies the spaces between fantasy and reality, ideal and corporeal.

In one such story, in 1775, the renowned Scottish anatomist William Hunter was commissioned by his friend Martin van Butchell (1735–1814), an eccentric dentist, to embalm his deceased wife, Maria. Van Butchell displayed his preserved spouse, along with her taxidermied pet parrot, in his living room, where

WITH·WONDERFUL·SKILL,
HE·CARVED·A·FIGURE,
BRILLIANTLY,·OUT·OF·SNOW-
WHITE·IVORY,·NO·MORTAL
WOMAN,·AND·FELL·IN·LOVE
WITH·HIS·OWN·CREATION.·THE
FEATURES·ARE·THOSE·OF·A
REAL·GIRL,·WHO,·YOU·MIGHT
THINK,·LIVED,·AND·WISHED
TO·MOVE,·IF·MODESTY·DID
NOT·FORBID·IT.·INDEED,·ART
HIDES·HIS·ART.·HE·MARVELS:
AND·PASSION,·FOR·THIS
BODILY·IMAGE,·CONSUMES
HIS·HEART.·OFTEN,·HE·RUNS
HIS·HANDS·OVER·THE·WORK,
TEMPTED·AS·TO·WHETHER
IT·IS·FLESH·OR·IVORY,·NOT
ADMITTING·IT·TO·BE·IVORY.
HE·KISSES·IT·AND·THINKS·HIS
KISSES·ARE·RETURNED;·AND
SPEAKS·TO·IT;·AND·HOLDS
IT,·AND·IMAGINES·THAT·HIS
FINGERS·PRESS·INTO·THE
LIMBS,·AND·IS·AFRAID·LEST

An extract from 'Orpheus sings: Pygmalion and the statue' in Ovid's Metamorphoses, Book X (8 CE). Pygmalion, 'offended by the failings that nature gave the female heart', sculpts his ideal woman.

BRUISES APPEAR FROM THE PRESSURE. NOW HE ADDRESSES IT WITH COMPLIMENTS, NOW BRINGS IT GIFTS THAT PLEASE GIRLS, SHELLS AND POLISHED PEBBLES, LITTLE BIRDS, AND MANY-COLOURED FLOWERS, LILIES AND TINTED BEADS, AND THE HELIADES'S AMBER TEARS, THAT DRIP FROM THE TREES. HE DRESSES THE BODY, ALSO, IN CLOTHING; PLACES RINGS ON THE FINGERS; PLACES A LONG NECKLACE ROUND ITS NECK; PEARLS HANG FROM THE EARS, AND CINCTURES ROUND THE BREASTS. ALL ARE FITTING: BUT IT APPEARS NO LESS LOVELY, NAKED. HE ARRANGES THE STATUE ON A BED ON WHICH CLOTHS DYED WITH TYRIAN MUREX ARE SPREAD, AND CALLS IT HIS BEDFELLOW, AND RESTS ITS NECK AGAINST SOFT DOWN, AS IF IT COULD FEEL.

Working in Marble or The Artist Sculpting Tanagra (1890) by Jean-Léon Gérôme. A self-portrait of the artist sculpting a replica of a beautiful model.

[4] Pygmalion and Galatea (c. 1890) by Jean-Léon Gérôme. Pygmalion kisses his creation and, with Venus's intervention, brings her to life.

fig. 93 The embalmed body of Eva Peron, photographed in 1952. Over the years, it was stolen, hidden, buried in secret, then exhumed and possibly sexually violated.

fig. 94 The beautiful Cuban-American woman Maria Elena Milagro de Hoyos, the love of Carl Tanzler's life, who he believed he had first met in a dream.

she could be viewed by appointment until his second wife forced him to get rid of her. She was then donated to William's brother, John, for display in his famous medical museum in London's Lincoln's Inn Fields, until she was destroyed in 1941 by a German firebomb.

A similar tale involves Esther Lachmann (1819–84), known as 'La Païva' and extolled by society chronicler Count Horace de Viel-Castel as 'the queen of kept women, the sovereign of her race'. Possibly the inspiration for Césarine in Alexandre Dumas's play *La Femme de Claude* (1873), she was born impoverished in Russia and went on to become one of the most successful and infamous courtesans in Paris. After La Païva died in 1884 aged sixty-four, her husband, Count Guido Henckel von Donnersmarck, had her body preserved in embalming fluid and kept it in the attic of his castle, where he nightly cried over it before it was discovered—and one assumes, disposed of—by his second wife.

A more recent case took place in Florida in the 1930s, when German-born radiologist Carl Tanzler (1877–1952) met Maria Elena Milagro de Hoyos (1909–1931),

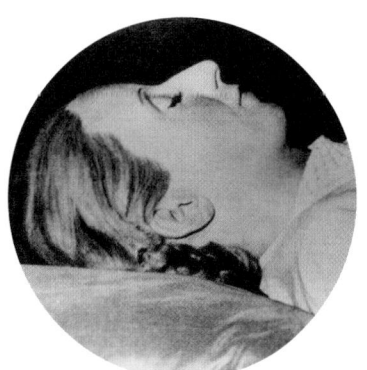

fig. 93

fig. 94

fig. 95

fig. 95 Effigy of Maria Elena Milagro de Hoyos, created by Carl Tanzler and kept at his home after she died of tuberculosis at the age of twenty-two. He lived with the doll until he died in 1952.

a young and beautiful patient of Cuban-American ancestry, and recognized her from a dream in which she had been introduced as the love of his life. Shortly afterwards, Tanzler attempted to heal her of tuberculosis with his self-professed medical knowledge, but to no avail. It is difficult to determine the truth about what happened after de Hoyos died in 1931. According to one version, Tanzler paid for her funeral and commissioned a mausoleum to contain her remains. Some eighteen months later, he transported her body on a toy wagon to a workshop he had set up inside an old aeroplane. He stuffed the corpse's cavities to preserve its form, fitted glass eyes, dressed it and, when decay made it necessary, added a wig made of de Hoyos's real hair. Tanzler doused the body with disinfectants, perfume and preservative fluids, and even installed a curtain to protect her modesty as they shared his bed. In 1940, when the body was discovered and removed, some claimed it contained a paper tube in its vagina to facilitate sexual intercourse. Tanzler then used a death mask as the basis for a life-sized effigy of de Hoyos with which he lived until his death in 1952.

From the annals of art history comes another story, this one of the Austrian expressionist painter Oskar Kokoschka (1886–1980) and the life-sized fetish he made of his former lover, Alma Mahler (1879–1964), the widow of composer Gustav Mahler, and one of the most pursued and celebrated women in Vienna. Having met in 1912 when he was twenty-six and she thirty-three, their fraught, three-year love affair became both sexual obsession and muse to the painter. Throughout, Mahler refused to publicly acknowledge their relationship and continued to carry on other affairs; she also aborted a pregnancy against his wishes. While Kokoschka was, in part due to her encouragement, fighting as a volunteer soldier in World War I, she married the famous German architect and founder of the Bauhaus art and design school, Walter Gropius (1883–1969).

Three years after the end of their relationship, Kokoschka commissioned toy maker Hermione Moos to create an effigy of his former lover to exacting specifications: 'Can the mouth be opened? And are there teeth and a tongue inside? I hope so!' When the much-anticipated doll finally arrived in 1919, Kokoschka was, unsurprisingly, deeply disappointed. Nevertheless, he painted and photographed the doll many times, and even took it out as his companion to the

fig. 96 Austrian socialite Alma Mahler (née Schindler), photographed c. 1900, two years before her marriage to Gustav Mahler.

fig. 96 *fig. 97* *fig. 98*

theatre and to restaurants. This public aspect suggests an element of revenge, a blow of humiliation intended to the woman who could not be controlled or possessed, and who refused to openly acknowledge their relationship. The doll also seems at least partly to have been conceived as an art project, given that Kokoschka published his own letters to Moos. Eventually, Kokoschka's doll was ceremonially doused in red wine and beheaded at a party. The following day, in the words of the painter, 'the dustcart came in the grey light of dawn, and carried away the dream of Eurydice's return. The doll was an image of spent love that no Pygmalion could bring to life.'

Of course, these examples are, in a sense, precursors to the much more prosaic but no less bizarre phenomenon of dolls crafted to serve as romantic and sexual companions for men who, perhaps like Pygmalion himself, find themselves chronically disappointed in the gulf between their notion of the ideal and the real in womanhood. This idea has a long history, but finds its real-world culmination in the phenomenon of the 'Real Doll', those hyper-idealized and extremely life-like dolls of women created to serve as companions and sexual partners.

fig. 97 Austrian expressionist painter Oskar Kokoschka in 1921.

fig. 98 Hermione Moos, toy maker and possibly also dressmaker for Alma Mahler.

THIS PAGE & OVERLEAF
*Gelatin silver print (1919) showing
Oskar Kokoschka's effigy of Alma Mahler
seated on a chair attended by her creator,
Hermione Moos, and posed as an odalisque.*

YESTERDAY·I·SENT·A
LIFE-SIZE·DRAWING·OF
MY·BELOVED·AND·I·ASK
YOU·TO·COPY·THIS·MOST
CAREFULLY·AND·TRANSFORM
IT·INTO·REALITY·WITH·THE
APPLICATION·OF·ALL·YOUR
PATIENCE·AND·FEELING.·PAY
SPECIAL·ATTENTION·TO·THE
DIMENSIONS·OF·THE·HEAD
AND·NECK,·TO·THE·RIBCAGE,
THE·RUMP·AND·THE·LIMBS.
AND·TAKE·TO·HEART·THE
CONTOURS·OF·THE·BODY,
E.G.·THE·LINE·OF·THE·NECK
TO·THE·BACK,·THE·CURVE
OF·THE·BELLY…THE·FIGURE
MUST·NOT·STAND!·THE·POINT
OF·ALL·THIS·FOR·ME·IS·AN
EXPERIENCE·WHICH·I·MUST
BE·ABLE·TO·EMBRACE!

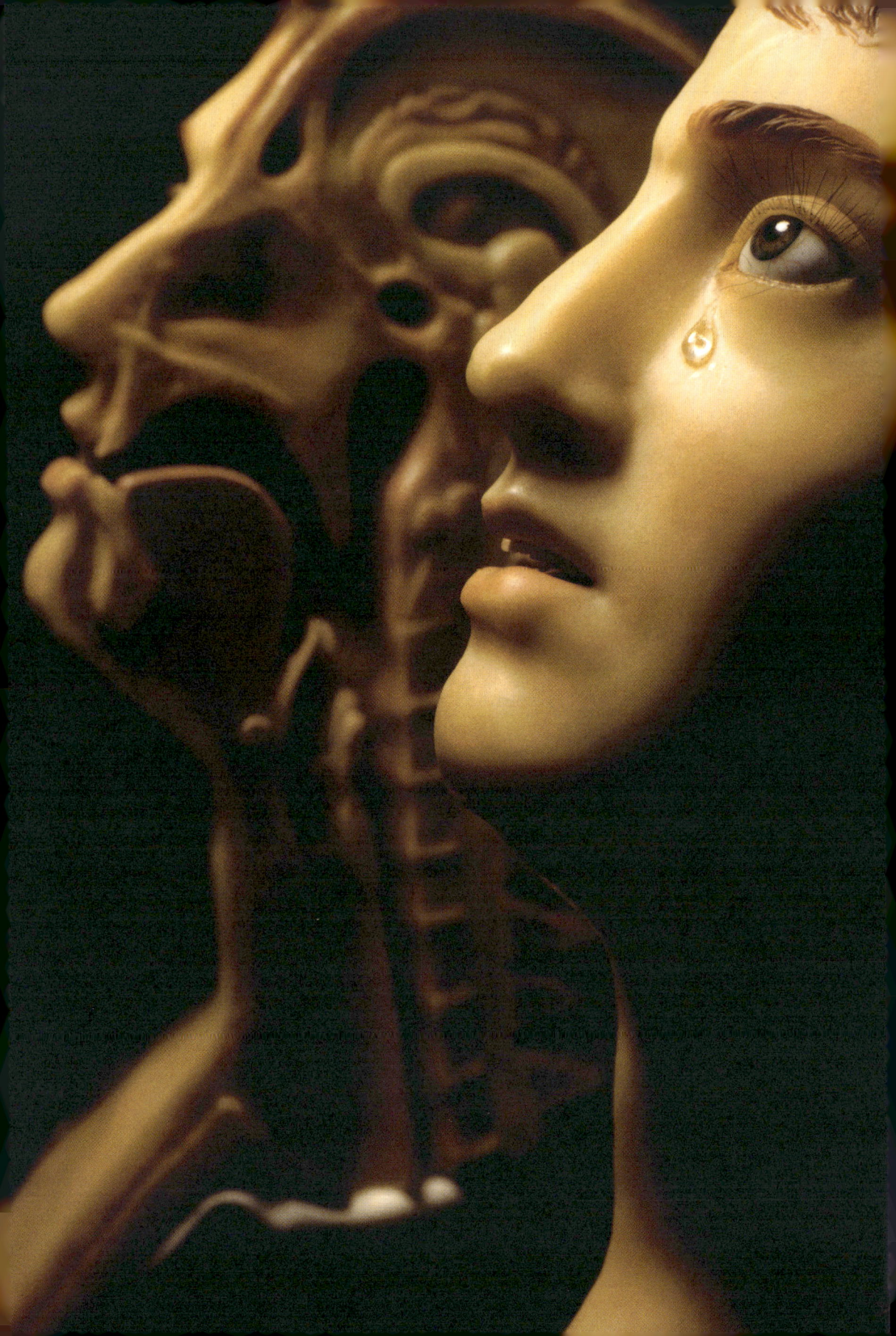

CHAPTER FIVE

VENUS, THE UNCANNY, AND THE GHOST IN THE MACHINE

(201)

PREVIOUS
Femme à la larne
(Woman with a teardrop), created in the late eighteenth century by André Pierre Pinson (1746–1828) for the private cabinet of the Duc D'Orléans.

Many people find the Anatomical Venus disquieting. Grotesque and beautiful, spectacle and teaching tool, seemingly both dead and alive, she tends to elicit a strong emotional engagement and intellectual uncertainty. This uncertainty, and the feelings of uneasiness it provokes, was described by Ernst Jentsch (1867–1919) in his essay 'On the Psychology of the Uncanny' (1906), and most famously elaborated upon by Sigmund Freud (1956–1939) in his essay 'The Uncanny' (1919), in which he quoted German philosopher F. W. J. Schelling's definition

fig. 99

fig. 99 Early-twentieth-century, life-sized wax fashion mannequin, probably of American origin, with glass eyes and human hair.

of the word: 'Uncanny is what one calls everything that was meant to remain secret and hidden, and has come into the open.'

The sensation of uneasiness described by Freud is felt when we encounter something that seems to confirm beliefs we held as children but were taught could not be true—such as that our toys might come to life. It confirms the discarded beliefs of our ancestors—the existence of revenants (the returned dead) or ghosts. Such encounters give us a sense of the *unheimlich*, or 'familiar unfamiliar': the uncanny. Our most basic notions of reality are challenged, causing us to question all that we take for granted—such as the non-existence of the supernatural or, as with automatons and life-like waxworks, the impossibility of seeming both dead and alive—and hurtling us into a world of atavistic archetypes. If a doll can come to life, the assumptions on which we built our understanding of the world are completely eroded. For some people this is a horrifying experience, and for others a pleasurable one.

VENUS, THE UNCANNY, AND THE GHOST IN THE MACHINE

The uncanny is related to the idea of the sublime, as articulated by the statesman and philosopher Edmund Burke (1729–97) in his treatise on aesthetics, *A Philosophical Enquiry into the Origin of Our Ideas of the Sublime and Beautiful* (1757):

> *Whatever is fitted in any sort to excite the ideas of pain and danger, that is to say, whatever is in any sort terrible, or is conversant about terrible objects, or operates in a manner analogous to terror...that is, it is productive of the strongest emotion*

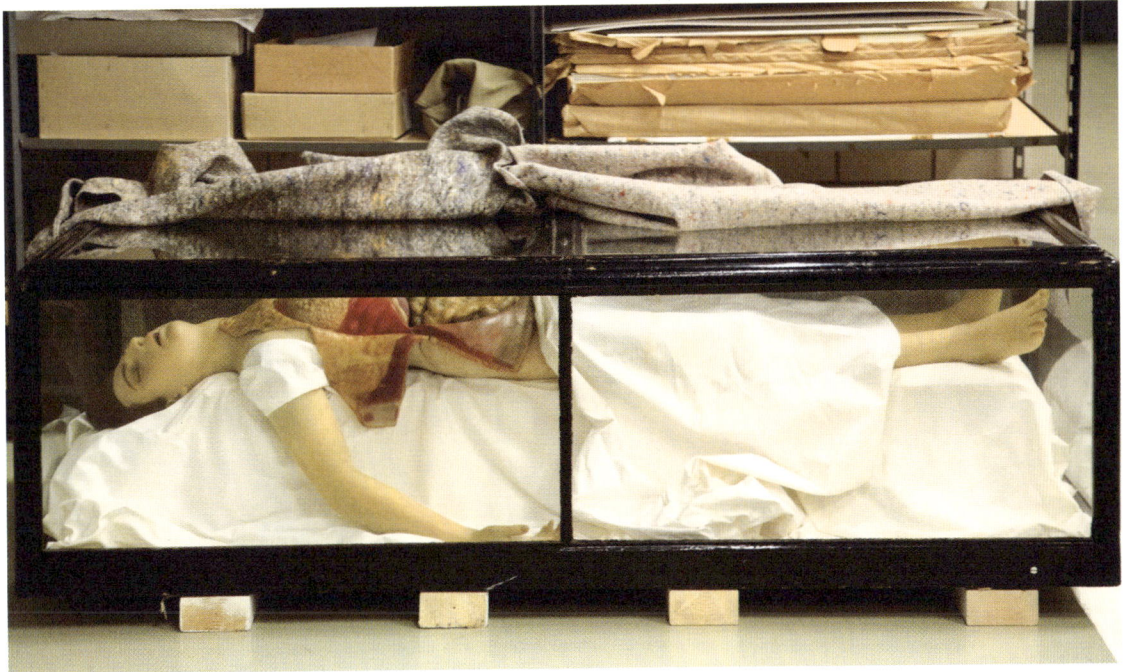

fig. 100 Wax Anatomical Venus (c. 1900) from the workshop of Rudolf Pohl, Dresden, Germany, photographed in storage at the Deutsches Hygiene-Museum, also in Dresden.

fig. 100

> *which the mind is capable of feeling....When danger or pain press too nearly, they are incapable of giving any delight, and are simply terrible; but at certain distances, and with certain modifications, they may be, and they are, delightful, as we every day experience.*

The uncanny is also related to the notion of the abject as described by the French-Bulgarian psychoanalyst Julia Kristeva (b. 1941), which 'refers to a human reaction (horror, vomit) to a threatened breakdown in meaning caused by the loss of the distinction between subject and object, or between self and other. The primary example for what causes such a reaction is the corpse (which traumatically reminds us of our own materiality).'

American cultural critic Terry Castle (b. 1953) argues that the uncanny is a product of the Enlightenment, which, via its zealous 'impulse to systematize and regulate', was responsible 'for that "estranging the real"...which is so integral a

OVERLEAF
An illustrated interpretation of the graph first created in 1970 by robotics professor Masahiro Mori, expressing his assertion that, up to a point, the more human an object or being appears, the more affinity we feel towards it. However, objects or beings such as mannequins or zombies, which are very similar to us while remaining distinctly other, provoke unease or revulsion, falling into a zone known as the 'uncanny valley'.

The Uncanny Valley
HUMANOID ROBOT
INDUSTRIAL ROBOT
MANNEQUIN
AFFINITY
HUMAN LIKENESS
TRILOBITE
TEDDY BEAR
BLOW-UP DOLL
VENTRILOQUI
DUMMY
CORPSE
MOVING
STILL
THE UNCANNY VALLEY

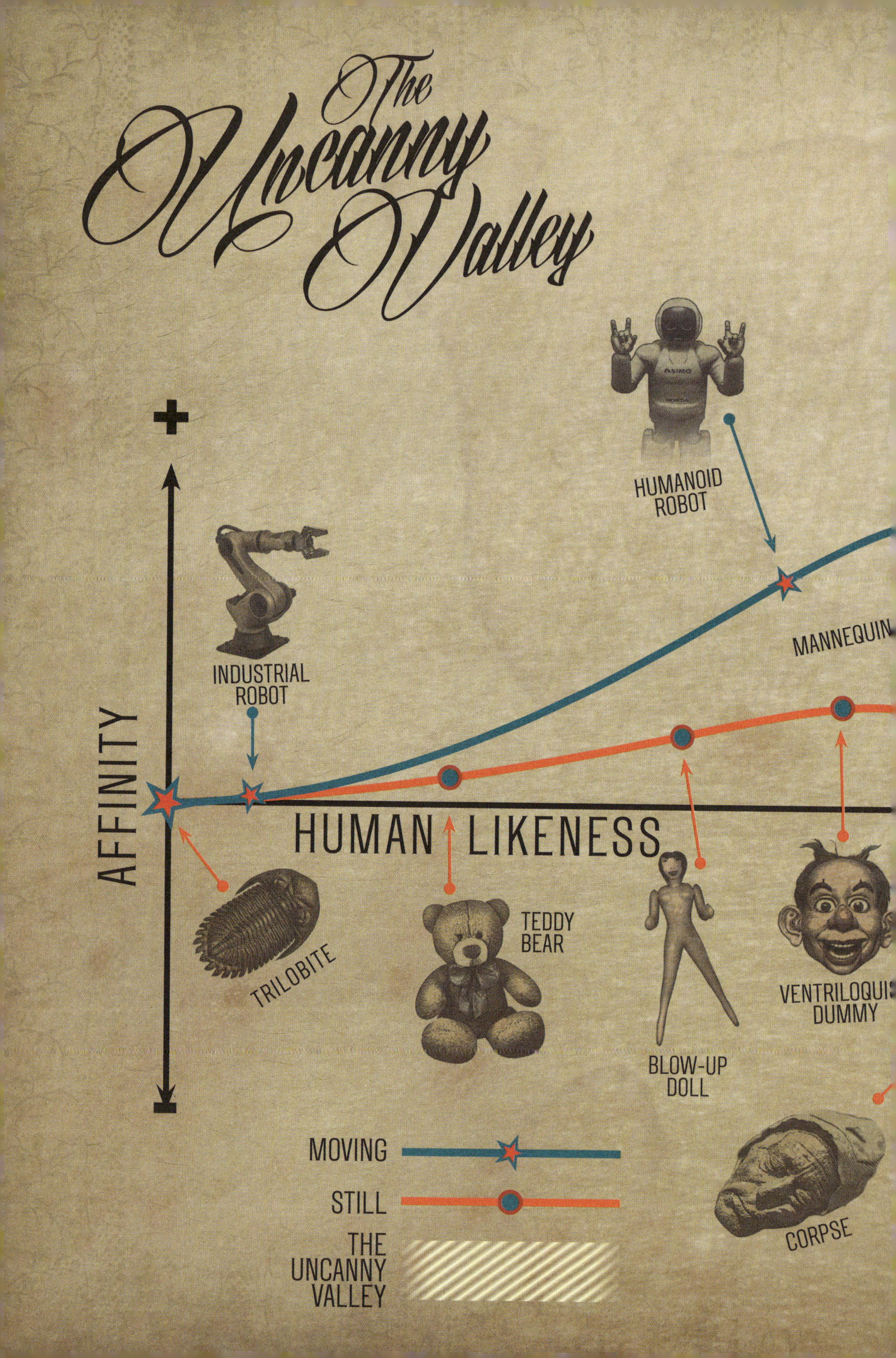

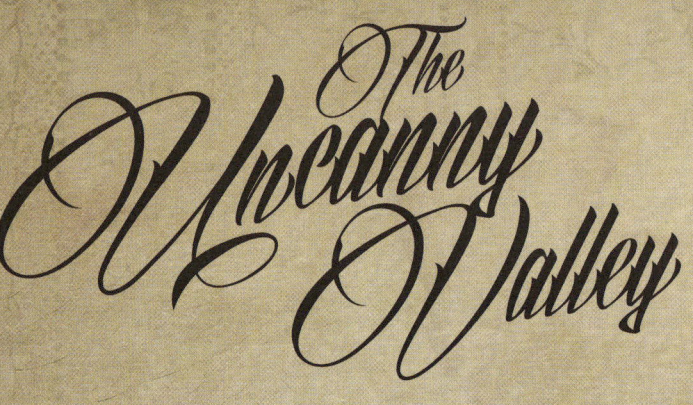

UNCANNY IS WHAT ONE CALLS EVERYTHING THAT WAS MEANT TO REMAIN SECRET AND HIDDEN AND HAS COME INTO THE OPEN.

PREVIOUS

F. W. J. Schelling, as quoted in 'The Uncanny', an essay by Sigmund Freud, first published in Imago, Bd. V, in 1919.

part of modernity.' It is, she argues, 'only when the "marvellous" is dislodged and "sober truth" elevated in its stead' that the uncanny can exist. Paradoxically, it is precisely our attempt to dispel superstition and create a rational, controllable universe that gives rise to the spectres that haunt the modern Western outlook. In 1865, the Irish historian William Lecky (1838–1903) observed the following:

> *Certainly no change in the history of the last 300 years is more striking, or suggestive of more curious enquiries, than that which has taken place in the estimate of the miraculous…a few centuries ago, there was no solution to which the mind of man turned more readily in every perplexity. A miraculous account was then universally*

fig. 101

fig. 102

figs 101–104
A selection of photographs by German artist Hans Bellmer (1902–75) that appeared in a collection of photographs published in 1935, entitled La Poupée, Seconde Partie (The Doll, Part II). Bellmer first produced a series of modular, life-sized, pubescent female dolls that he posed and photographed in 1934, publishing ten in a collection entitled Die Puppe (The Doll). The second collection of photographs features a more flexible doll, which he photographed in various provocative scenarios.

accepted as perfectly credible, probable, and ordinary…. The powers of light and the powers of darkness were regarded as visibly struggling for the mastery. Saintly miracles, supernatural cures, startling judgments, visions, prophecies, and prodigies of every order, attested the activity of the one, while witchcraft and magic, with all their attendant horrors, were the visible manifestations of the latter…

Indeed, until Enlightenment ideas triumphed, the supernatural was an important and commonly accepted way of understanding the world. For many living in a largely disenchanted, post-Cartesian world, the soul has been vanquished and the body and mind torn asunder. British philosopher Gilbert Ryle (1900–76) rejected such Cartesian materialism in his book *The Concept of Mind* (1949), referring to it with 'deliberate abusiveness' as 'the dogma of the Ghost in the Machine'. Yet Descartes learned the principles of materialist rationality from an angel who visited him in his dreams. Even the most materialist of fantasies, in which Nature and God are trumped by man-made machine, often personify that machine as an automaton, implying a link between human appearance and the possession of an independent mind or spirit. This impulse to relate to a sense of agency in a machine appears in the decadent novel *À Rebours* ('Against Nature') by J. K. Huysmans (1848–1907), in which the aristocratic protagonist Jean Des Esseintes describes a new French locomotive as if it were a woman:

> ...the Crampton, an adorable blonde, shrill voiced, slender-waisted, with her glittering corset of polished brass...the perfection of whose charms is almost terrifying when, stiffening her muscles of steel, pouring the sweat of steam down her flanks, she sets revolving a puissant circle of her elegant wheels...

fig. 103

fig. 104

The materialist worldview privileges mind over body, the quantifiable over the intuited, theory over experience. The Anatomical Venus can be seen as part of an Enlightenment project to impose a sense of order and control on that which will always defy our understanding. To anatomize is, after all, to single out, to separate from its system, to analyse each thing separately. Yet this approach can obscure knowledge of equal importance: an understanding of the system as a whole, connections, and the relationship between things. And it brings us no closer to understanding the mysteries of life and death at the core of our own fascination. As Immanuel Kant (1724–1804) stated in the opening sentence of one of the most iconic texts of Enlightenment thinking, *Critique of Pure Reason* (1781):

> Human reason has this peculiar fate, that in one species of its knowledge it is burdened by questions which, as prescribed by the very nature of reason itself, it is not able to ignore, but which, as transcending all its powers, it is also not able to answer.

DOUBLES, DANCING DOLLS AND AUTOMATA, WAXWORK FIGURES, ALTER EGOS AND 'MIRROR SELVES', SPECTRAL EMANATIONS... WHAT MAKES THEM UNCANNY IS PRECISELY THE WAY THEY SUBVERT THE DISTINCTION BETWEEN THE REAL AND THE PHANTASMATIC — PLUNGING US INSTANTLY, AND VERTIGINOUSLY, INTO THE HAG-RIDDLED WORLD OF THE UNCONSCIOUS.

Excerpt from The Female Thermometer (1995) by Terry Castle in which she expounds on the nature of the uncanny.

An unnerving piece from the series Tragic Anatomies *(1996) by Jake and Dinos Chapman. The installation featured mannequins in a variety of unsettling poses.*

Atheists have argued that scientific knowledge can produce a sense of awe or wonder about a world in which God is 'dead'. But even the most atheistic can feel as though science does not have all the answers. In 1845 French physician and psychiatrist Alexandre Jacques François Brière de Boismont (1797–1881) observed:

> ...the feeling of the unknown to which man attaches himself, and from which arises the want of something to believe, a love of the marvellous, a desire for knowledge, and a craving after excitement, is itself only a weakened condition of the religious sentiment.

The non-religious among us continue to seek the transcendent emotional thrills of the miraculous and the sacred, finding them in manifestations of the sublime and the uncanny, the numinosity of objects, and the ecstatic release of sex and drugs. We experience wonder at fairgrounds, theatres, cinemas, and museums,

fig. 105

fig. 106

figs 105–108 Four photographs of sex dolls by Stacy Leigh, from her series Average Americans. The dolls are posed in sets of Leigh's own creation. She says, 'I think that there is something that happens when you see a hyper-realistic, life-size doll...in person or in a photo, there is a feeling. Something that looks so alive, yet it is so still. It's uncomfortable. You might feel repulsed and empathetic at the same time.'

and find a way to shed our personal boundaries at concerts and sporting events. We continue to seek the psychological reassurances of confession in the form of the 'talking cure' of psychoanalysis, in online confessionals, and in tell-all memoirs. We continue to demonize and mortify the flesh, as did the saints, with rigid or punishing diets and extreme workouts.

Medicine, too, sublimates the magical, in ways formally and functionally similar to religion. Ex-votos and the cult of the saints represent our appeal to religion for help against death and disease, whereas medicine seeks science's intervention. The mysterious efficacy of the placebo calls into question a neat split between mind and body, while drawing attention to the theatrical, persuasive power of the white coat itself. If, as Orson Welles (1915–85) attests in the documentary F is for Fake (1973), experts are the oracles of the modern age, what expert is more powerful than the one with the power to vanquish death—for a time, at least—on our behalf? Doctors and priests both have their roots in the shaman, healer of soul and body, and tribal medicine men use the power of spectacle to effect change. What is medicine, after all, but the latest (and most successful) in a long line of strategies intended to cheat death and ensure personal survival, regardless of fate, destiny, or God's will?

In fashioning the ideal woman inside and out, Enlightenment man's ambitious aims included civilizing and educating a new, sophisticated breed of citizen; solving the problems of dissection; and both understanding and actually embodying the mind of God. Yet, centuries later, many of us no longer feel so confident in the possibility of penetrating the secrets of a benign Mother Nature, of an invitingly passive Anatomical Venus who will lead us to discover the universe. An updated feminine myth of nature and the universe might look more like scientist James Lovelock's conception of 'Gaia'—a complex, self-regulating system of life and the ecosphere: crucially, an indifferent force, much bigger than humankind, and one that will outlast it.

The Anatomical Venus, then, is an object lesson: a reminder that science and its artefacts are never truly neutral. Science does not simply uncover truth; it is also a culturally constructed, normative activity that reflects the ideals of its time and sublimates human drives, such as desiring, collecting, and finding

fig. 107

fig. 108

meaning. The Anatomical Venus cautions us that the omniscient dreams of the Enlightenment were never fully realized and cannot be, so long as our psychologies remain as they are. Reality defies the categories we employ to organize knowledge, and extends well beyond the limits of human perception.

Our attraction to the Anatomical Venus appears to suggest, by its very existence, that archaic ways of thinking could be more relevant than we realise; that the spiritual or supernatural is not nonsense and that the non-material realm should not be relegated to the mind. Perhaps she causes us to unify the psychological drives Eros, to love, and Thanatos, to die. These divisions are products of our own contemporary perspective—unique to our own time—and say more about ourselves than they do about her.

Perhaps the draw of the Anatomical Venus comes from an unspoken, intuited resolution of our own divided nature, an unconscious recognition of another avenue abandoned, in which beauty and science, religion and medicine, soul and body might be one. And maybe she tells us that—as people once believed, but we currently discredit—the microcosm really does reflect the macrocosm, and one really can find the entire world in a grain of sand, or the entire universe in a single object.

OVERLEAF
A modern interpretation of the Anatomical Venus, created from a photograph by Koen Hauser for the Museum Boerhaave's 'Amazing Models' exhibition in Leiden, Netherlands (2013–14).

PLACES·OF·INTEREST

[AUSTRALIA]

HARRY BROOKES ALLEN MUSEUM
harrybrookesallenmuseum.mdhs.unimelb.edu.au
University of Melbourne, Medical Building, Parkville
VIC 3010 | *Houses around 12,000 artefacts with histories stretching back to the 1800s.*

[AUSTRIA]

JOSEPHINUM | josephinum.ac.at | Währinger Straße 25, 1090 Vienna | *Best known for its collection of 1,192 anatomical and obstetric wax models.*

PATHOLOGISCH-ANATOMISCHE SAMMLUNG IM NARRENTURM | nhm-wien.ac.at/forschung/anthropologie | Hof 6, Spitalgasse 2, A-1090 Vienna *Founded in 1796 and housed in the so-called Madhouse Tower ('Narrenturm').*

PRATERMUSEUM | Oswald-Thomas-Platz 1, 1020 Vienna | *Houses a collection of artefacts relating to The Prater, an amusement park founded in 1766.*

[BELGIUM]

MUSÉE DE LA MÉDECINE DU BRUXELLES www.museemedecine.be | Campus Erasme – Place Facultaire, Lenniksebaan 808, 1070 Anderlecht *Includes waxes once exhibited by the Musée Fujy, a competitor of the Spitzner Museum.*

[COLOMBIA]

MUSEO DE HISTORIA DE LA MEDICINA | Carrera 30 calle 45-05, Universidad Nacional de Colombia, Sede Bogotá | *A collection of wax dermatological moulages created by Professor Manuel José Silva.*

[DENMARK]

MEDICAL MUSEION | www.museion.ku.dk | Bredgade 62, DK-1260 Copenhagen | *A collection of over 250,000 objects related to Danish medicine, some with origins in the early seventeenth century.*

[FRANCE]

MUSÉE D'ANATOMIE DE MONTPELLIER umontpellier.fr/universite/patrimoine/musees 2 rue de l'École de Médecine, 34000 Montpellier *Waxes from the workshop at La Specola and the Sleeping Venus once displayed at the Musée Spitzner.*

MUSÉE D'HISTOIRE DE LA MÉDECINE | 12 rue de l'Ecole de Médecine, 75006 Paris | *This collection, among the oldest in Europe, includes a life-sized wooden dissectible model crafted by Felice Fontana.*

MUSÉE TESTUT LATARJET D'ANATOMIE ET D'HISTOIRE NATURELLE MÉDICALE | museetl.univ-lyon1.fr | 8 avenue Rockefeller, 69008 Lyon *Home to a substantial collection of wax anatomical models.*

[GERMANY]

DEUTSCHES HYGIENE-MUSEUM | dhmd.de Lingnerplatz 1, 01069 Dresden | *Best known for its life-sized 'Transparent Man' (c. 1930). Also displays a collection of popular anatomical waxworks.*

MÜNCHNER STADTMUSEUM | muenchner-stadtmuseum.de | Sankt-Jakobs-Platz 1, 80331 München | *Houses a dissectible Anatomical Venus made in the early 1930s, among other popular waxworks.*

[GREECE]

MOULAGE MUSEUM OF ANDREAS SYGROS HOSPITAL www.universeum2015.uoa.gr | Ionos Dragoumi 5, 16121 Athens | *One of the largest collections of moulages in the world.*

[HUNGARY]

SEMMELWEIS MUSEUM | semmelweismuseum.hu Apród utca 1–3, H-1013 Budapest | *A rich medical and artistic collection featuring an Anatomical Venus from the workshop at La Specola.*

[ITALY]

CIMITERO DELLE FONTANELLE | cimiterofontanelle.com | Via Fontanelle 80, 80136 Napoli | *The centre of the Neapolitan Skull Cult, whose devotees adopt and care for the unidentified skulls of the dead.*

COLLEZIONE DELLE CERE ANATOMICHE DI CLEMENTE SUSINI | pacs.unica.it/cere | Piazza Arsenale 1, 09124 Cagliari | *A substantial collection including some of Clemente Susini's finest late-career anatomical waxes.*

MUSEO DELLE CERE ANATOMICHE 'LUIGI CATTANEO' | museocereanatomiche.it Via Irnerio 48, 40126 Bologna | *Anatomical waxes by Clemente Susini and anatomical models by Giuseppe Astorri and Cesare Bettini.*

MUSEO DI PALAZZO POGGI | museopalazzopoggi.unibo.it | Via Zamboni 33, 40126 Bologna | *Famous for Clemente Susini's dissectible Venerina and anatomical waxes by Ercole Lelli and Anna Morandi Manzolini.*

MUSEO DI ANATOMIA PATOLOGICA DELL'UNIVERSITÀ DEGLI STUDI DI FIRENZE | Viale Morgagni 85, 50134 Florence | *Features pathological waxes by Luigi Calamai, an artist who worked at La Specola.*

MUSEO DI ANATOMIA, UNIVERSITÀ DI PAVIA | musei.unipv.it/musei/2_musei_6_AN.html | via Forlanini 8, 27100 Pavia | *This gorgeous museum displays a particularly fine Anatomical Venus, created in 1794 by Clemente Susini.*

MUSEO DI STORIA NATURALE 'LA SPECOLA' msn.unifi.it | Via Romana 17, 50125 Firenze *La Specola's famed wax workshop produced the finest of all anatomical wax models and Anatomical Venuses.*

TEATRO ANATOMICO DI BOLOGNA | www.archiginnasio.it/teatro.htm | Piazza Galvani 1, 40124 Bologna *Features Ercole Lelli's 1734 wooden écorché figures. He went on to create the wax models now housed at Museo di Palazzo Poggi.*

[MEXICO]

PALACIO DE LA ESCUELA DE MEDICINA pem.facmed.unam.mx | Brasil #33, Cuauhtémoc, 06010 Mexico City | *Wonderful collection of nineteenth-century wax moulages and models, housed in the former Palace of the Inquisition.*

[NETHERLANDS]

MUSEUM BOERHAAVE | museumboerhaave.nl Lange St. Agnietenstraat 10, 2312 WC Leiden *Features, among other treasures, a reconstruction of Leiden's seventeenth-century anatomical theatre.*

MUSEUM VROLIK | Meibergdreef 15, J0-130, 1105 AZ Amsterdam | *Began as the nineteenth-century private collection of Gerard and Willem Vrolik, professors of anatomy.*

[RUSSIA]

KUNSTKAMERA | kunstkamera.ru | 3 Universitetskaya Embankment, St Petersburg 199034 | *The first public museum in Russia, best known today for holding most of the extant work of Frederik Ruysch.*

[SPAIN]

MUSEO DE ANATOMÍA JAVIER PUERTA | pendiente demigracion.ucm.es/info/museoana | Ciudad Universitaria s/n, 28040 Madrid | *Founded in the nineteenth century. Includes a life-sized wooden anatomized Adam and Eve and a life-sized wax skeleton.*

[SWEDEN]

MUSEUM GUSTAVIANUM | www.gustavianum.uu.se Akademigatan 3, 763 10 Uppsala | *A collection tracing back to Carl Linnaeus.*

[SWITZERLAND]

MOULAGENMUSEUM | Haldenbachstrasse 14, 8006 Zürich | *A vast and renowned collection of dermatological wax moulages.*

[UK]

THE GORDON MUSEUM OF PATHOLOGY kcl.ac.uk/gordon | Hodgkin Building, Guy's Campus, King's College London | *A noteworthy collation of nineteenth-century wax models and moulages by British modeller-in-residence Joseph Towne.*

NATIONAL FAIRGROUND ARCHIVE | sheffield.ac.uk/nfa | Western Bank Library, University of Sheffield, Sheffield | *An incredible source of material relating to the history of popular entertainment in the UK, beginning in the nineteenth century.*

THE SCIENCE MUSEUM, LONDON | sciencemuseum.org.uk/broughttolife | Exhibition Rd, London *The history of medicine display includes a half-size Anatomical Venus that was almost certainly the scale model for the final pieces at La Specola.*

MADAME TUSSAUDS | madametussauds.co.uk Marylebone Rd, London | *Home of the Sleeping Beauty, a waxwork woman that appears to be breathing, slumbering on a chaise.*

THE WELLCOME COLLECTION | wellcomecollection.org 183 Euston Rd, London | *Henry Wellcome's breathtaking collection of objects related to health, medicine, and the human condition.*

[USA]

THE LIBRARY COMPANY OF PHILADELPHIA librarycompany.org | 1314 Locust St, PA 19107 *Noteworthy for housing the Bill Helfand collection of medical ephemera, including popular museum guidebooks and posters.*

MORBID ANATOMY MUSEUM | morbidanatomymuseum.org | 424A Third Avenue, Brooklyn, NY 11215 *Exhibitions and extensive research library relating to Anatomical Venuses, art, medicine, death, and culture.*

THE MÜTTER MUSEUM | muttermuseum.org 19 S 22nd Street, PA 19103 | *The most beloved medical museum in the United States, with a vast collection of wax models and moulages.*

NATIONAL MUSEUM OF HEALTH AND MEDICINE medicalmuseum.mil | 2500 Linden Lane, Silver Spring, MD 20910 | *Founded in 1862, the museum holds a large collection of anatomical specimens, models, and moulages.*

U.S. NATIONAL LIBRARY OF MEDICINE | nlm.nih.gov 8600 Rockville Pike, Bethesda, MD 20894 | *The world's largest biomedical library, a treasure trove of amazing images from anatomical history.*

SELECT·BIBLIOGRAPHY

[BOOKS AND PAMPHLETS]

ALLEY, R. (1981). *Catalogue of The Tate Gallery's Collection of Modern Art other than works by British Artists*. Tate Gallery in association with Sotheby Parke Bernet.
ALTICK, R. D. (1978) *The Shows of London*. Harvard University Press.
AMENDOLA, A. & PASTORINO, U. (2014). *Le cere vive di Clemente Susini*. FMR.
ARIÈS, P. (2008). *The Hour of Our Death: The Classic History of Western Attitudes Toward Death Over the Last One Thousand Years*. Vintage Books.
BARSANTI, G. & CHELAZZI, G. (2009). *The Museum of Natural History of the University of Florence, Vol. I: The Collections of La Specola: Zoology and Anatomical Waxes*. Firenze University Press.
BARTLETT, R. (2015). *Why Can the Dead Do Such Great Things? Saints and Worshippers from the Martyrs to the Reformation*. Princeton University Press.
BATAILLE, G. (1986). *Erotism: Death & Sensuality*. City Lights Books.
BLOOM, M. E. (2003). *Waxworks: A Cultural Obsession*. University of Minnesota Press.
BRAITHWAITE, P. (2001). *The Rise of Waxwork Shows from Pagan Times to World War One*. Waxworks Society.
BURMEISTER, M. R. (2000). *Popular Anatomical Museums in Nineteenth-Century England*. Ph.D. thesis, Rutgers University.
CALAMARI, B. & DIPASQUA, S. (2007). *Patron Saints*. Abrams.
CARLINO, A., COMAR, P. & CLAIR, J. (2008). *Figures du corps: Une leçon d'anatomie à l'Ecole des Beaux-Arts*. Beaux-Arts de Paris.
CASTLE, T. (1995). *The Female Thermometer: Eighteenth-Century Culture and the Invention of the Uncanny*. Oxford University Press.
CLOUSTON, W. A. (1876). *Literary Curiosities and Eccentricities: A Book of Anecdote Laconic Sayings, and Gems of Thought in Prose and Verse*. Ward, Lock and Tyler.
DANINOS, A. (ed.) (2012). *Waxing Eloquent: Italian Portraits in Wax*. Officina Libraria.
DOWNING, L. (2003). *Desiring the Dead: Necrophilia and Nineteenth Century French Literature*. European Humanities Research Centre, University of Oxford.
DUFFIN, J. (2014). *Medical Miracles: Doctors, Saints, and Healing in the Modern World*. Oxford University Press.
DÜRING, M. V., DIDI-HUBERMAN, G., POGGESI, M. & BAMBI, S. (2006). *Encyclopaedia Anatomica: A Complete Collection of Anatomical Waxes*. Taschen.
FAYE, B. & LEMIRE, M. (1990). *Artistes et mortels*. Chabaud.
FLEMING, J. V. (2013). *The Dark Side of the Enlightenment: Wizards, Alchemists, and Spiritual Seekers in the Age of Reason*. W. W. Norton & Company.
FREUD, S., MCLINTOCK, D. (trans.) & HAUGHTON, H (ed.). (2003). *The Uncanny*. Penguin Books.

GERCHOW, J. & BELTING, H. (2002). *Ebenbilder: Kopien von Körpern – Modelle des Menschen*. Hatje Cantz.
HUSTVEDT, A. (2006). *The Decadent Reader: Fiction, Fantasy and Perversion from Fin-de-Siècle France*. Zone Books.
JORDANOVA, L. J. (1993). *Sexual Visions: Images of Gender in Science and Medicine between the Eighteenth and Twentieth Centuries*. University of Wisconsin Press.
KEMP, M. & WALLACE, M. (2001). *Spectacular Bodies: The Art and Science of the Human Body from Leonardo to Now*. Hayward Gallery.
KRAFT-EBING, R. V. & KING, B. (1999). *Psychopathia Sexualis, with Especial Reference to Contrary Sexual Instinct: A Clinical-Forensic Study*. Bloat.
KÖNIG, H. & ORTENAU, E. (1962). *Panoptikum: vom Zauberbild zum Gaukelspiel der Wachsfiguren*. Isartal.
KOOIJMANS, L. (2011). *Death Defied: The Anatomy Lessons of Frederik Ruysch*. Brill.
KOUDOUNARIS, P. (2011). *The Empire of Death: A Cultural History of Ossuaries and Charnel Houses*. Thames & Hudson.
KRISTEVA, J. (2010). *Powers of Horror: An Essay on Abjection*. Columbia University Press.
MACCULLOCH, D. (2011). *Christianity: The First Three Thousand Years*. Penguin.
MAERKER, A. (2015). *Model Experts: Wax Anatomies and Enlightenment in Florence and Vienna, 1775-1815*. Manchester University Press.
MARTIN, J. R. (1991). *Baroque*. Penguin Books.
MARTY, M. E. (1986). *Modern American Religion, Volume 1: The Irony of It All, 1893-1919*. The University of Chicago Press.
MESSBARGER, R. M. (2010). *The Lady Anatomist: The Life and Work of Anna Morandi Manzolini*. The University of Chicago Press.
MUNRO, J. (2014). *Silent Partners: Artist and Mannequin from Function to Fetish*. Yale University Press.
NORTON, R. (1914). *Bernini, and Other Studies in the History of Art*. Macmillan.
OCKMAN, C. & SILVER, K. E. (2005). *Sarah Bernhardt: The Art of High Drama*. Yale University Press.
PANZANELLI, R., LOUGHRIDGE, M. & SCHLOSSER, J. V. (2008). *Ephemeral Bodies: Wax Sculpture and the Human Figure*. Getty Research Institute.
PHILIPS, D. (2012). *Fairground Attractions: A Genealogy of the Pleasure Ground*. Bloomsbury.
PILBEAM, P. M. (2003). *Madame Tussaud and the History of Waxworks*. Hambledon and London.
PURCELL, R. W. & GOULD S. J. (1992). *Finders, Keepers: Eight Collectors*. W. W. Norton & Company.
RIVA, A. (2007). *Flesh & Wax: The Clemente Susini's Anatomical Models in the University of Cagliari*. Ilisso.
ROYLE, N. (2008). *The Uncanny*. Manchester University Press.

SANDBERG, M. B. (2005). *Living Pictures, Missing Persons: Mannequins, Museums and Modernity*. Princeton University Press.

SAPPOL, M. (2004). *A Traffic of Dead Bodies: Anatomy and Embodied Social Identity in Nineteenth-Century America*. Princeton University Press.

SAPPOL, M. (2006). *Dream Anatomy*. National Library of Medicine.

SCHWARTZ, V. R. (2003). *Spectacular Realities: Early Mass Culture in Fin-de-Siècle Paris*. University of California Press.

SMITH, M. (2013). *The Erotic Doll: A Modern Fetish*. Yale University Press.

STEPHENS, E. (2013). *Anatomy as Spectacle: Public Exhibitions of the Body from 1700 to the Present*. Liverpool University Press.

TAYLOR, R. P. (1985). *The Death and Resurrection Show: From Shaman to Superstar*. Anthony Blond.

THOMAS, K. (2012). *Religion and the Decline of Magic: Studies in Popular Beliefs in Sixteenth- and Seventeenth-Century England*. Folio Society.

UNIVERSITÀ DI BOLOGNA (2001). *Guide to the Museo di Palazzo Poggi: Science and Art*. Editrice Compositori.

WARNER, M. (2008). *Phantasmagoria: Spirit Visions, Metaphors, and Media into the Twenty-First Century*. Oxford University Press.

WHITFORD, F. (1986). *Oskar Kokoschka: A Life*. Wiedenfeld and Nicolson.

[ARTICLES AND ESSAYS]

BALLESTRIERO, R. (2007). 'The history of ceroplastics/wax modelling' in RIVA, A. (ed.) *Flesh & Wax: The Clemente Susini's anatomical models in the University of Cagliari*. Ilisso, 17–34.

BALLESTIERO, R. (2010). 'Anatomical models and wax Venuses: art masterpieces or scientific craft works?'. *Journal of Anatomy*. 216 (2), 223–34.

BARNETT, R. (2008). 'Lost wax: medicine and spectacle in Enlightenment London'. *The Lancet*. 372 (9636), 366–67.

BATES, A. W. (2006). 'Dr Kahn's Museum: obscene anatomy in Victorian London' in *Journal of the Royal Society of Medicine*. 99 (12), 618–24.

BATES, A. W. (2006). Anatomical Venuses: the aesthetics of anatomical modelling in eighteenth and nineteenth-century Europe in PUSZTAI, J. (ed.) *40th International Congress on the History of Medicine: Proceedings*. Vesalius, Budapest. 1, 183–86.

CEGLIA, F. P. (2006). 'Rotten Corpses, a Disemboweled Woman, a Flayed Man. Images of the Body from the End of the 17th to the Beginning of the 19th Century. Florentine Wax Models in the First-hand Accounts of Visitors'. *Perspectives on Science*. 14 (4), 417–56.

CRASKE, M. (2010). '"Unwholesome" and "pornographic": a reassessment of the place of Rackstrow's Museum in the story of eighteenth-century anatomical collection and exhibition'. *Journal of the History of Collections*. jhc.oxfordjournals.org/content/23/1/75.short [November 19 2010].

DACOME, L. (2006). 'Waxworks and the performance of anatomy in mid-18th-century Italy'. *Endeavour*. 30 (1), 29–35.

DECKERS, R. (2013). '"La Scandalosa" in Naples: a veristic waxwork as memento mori and ethical challenge'. *The Oxford Art Journal*. 36 (1), 75–91.

DEER, L. (1977). 'Italian anatomical waxes in the Wellcome Collection: the missing link'. *Rivista di Storia delle Scienze Mediche e Naturali*. 20, 281–98.

GUERZONI, G. A. (2012). 'Use and Abuse of Beeswax in the Early Modern Age: Two Apologues and a Taste' in A. DANINOS (ed.) *Waxing Eloquent: Italian Portraits in Wax*. Officina Libraria, 43–60.

FERRARI, G. (1987). 'Public anatomy lessons and the carnival: the anatomy theatre of Bologna'. *Past & Present*. 117, 50–106.

HAVILAND, T. N. & PARRISH, L. L. C. (1970). 'A brief account of the use of wax models in the study of medicine'. *Journal of the History of Medicine and Allied Sciences*. 25 (1), 52–75.

HOFFMANN, K. A. (2006). 'Sleeping beauties in the fairground'. *Early Popular Visual Culture*. 4 (2), 139–59.

KNOEFEL, P. K. (1978). 'Florentine anatomical models in wax and wood'. *Medicina nei secoli*. 16 (3), 329–40.

LIGHTBOWN, R. W. (1964). 'Gaetano Giulio Zumbo – 1: the Florentine period'. *The Burlington Magazine*. 106 (740), 486–96.

LIGHTBOWN, R. W. (1964). 'Gaetano Giulio Zumbo – 2: Genoa and France'. *The Burlington Magazine*. 106 (741), 563–69.

MÄRKER, A. K. (2006). 'The anatomical models of La Specola: production, uses and reception'. *Nuncius*. 21 (2), 295–321.

MCISAAC, P. M. (2015) 'Castan's in context: introductory remarks on a bygone world in wax'. in exhibition catalogue, 'House of Wax', Morbid Anatomy Museum.

MESSBARGER, R. (2012). 'The re-birth of Venus in Florence's Royal Museum of Physics and Natural History'. *Journal of the History of Collections*. jhc.oxfordjournals.org/content/early/2012/05/16/jhc.fhs007 [May 16 2012].

PIOMBINO-MASCALI, D. & ZINK, A. (2011). 'The Fontanelle cemetery and the skull cult in contemporary Naples' in WIECZOREK, A. & ROSENDAHL, W. (ed.) *Schädelkult: Kopf und Schädel in der Kulturgeschichte des Menschen*. Schnell & Steiner, 263–65.

PULHAM, P. (2008). 'The eroticism of artificial flesh in Villiers de L'Isle Adam's *L'Eve Future*'. *19: Interdisciplinary Studies in the Long Nineteenth Century*, Issue 7. www.19.bbk.ac.uk/articles/abstract/10.16995/ntn.486/ [October 01 2008].

RADFORD, T. (2005). 'Secrets of the flesh peeled away'. *The Guardian*. 4 May.

RIVA, A., CONTI, G., SOLINAS, P. & LOY, F. (2010). 'The evolution of anatomical wax modelling in Italy from the 16th to early 19th centuries'. *Journal of Anatomy*. 216 (2), 209–22.

SAPPOL, M. (2004). Morbid curiosity: the decline and fall of the popular anatomical museum. *Common-Place*. 4 (2). www.common-place.org.

SCASCIAMACCHIA S., ET AL (2012). 'Plague epidemic in the Kingdom of Naples, 1656–1658'. *Emerging Infectious Diseases*. 18 (1), 186–88.

SIMUNI, F. (2009). 'Anatomie conturbanti' 'Perturbing anatomies' in BELLASI, P. & CORÀ, B. (ed.) *Bodies, Automata, Robots in Art, Science and Technology*. Mazzotta, 382–84.

PICTURE·CREDITS

T = TOP, B = BOTTOM, C = CENTRE, L = LEFT, R = RIGHT

[1] Wellcome Library, London [2] Deutsches Hygiene-Museum, Dresden. Photo David Brandt [4-5] Museo di Palazzo Poggi, Universita' di Bologna. Photo Joanna Ebenstein [6-7] Josephinum, Collections and History of Medicine, MedUni Vienna. Photo Joanna Ebenstein [8-9] Madame Tussauds Archives, London. Photo Joanna Ebenstein [10-11] (BOTH) Münchner Stadtmuseum, Sammlung Puppentheater / Schaustellerei, Munich [12] Wellcome Collection, Blythe House, London. Photo Joanna Ebenstein [14-15] Museo di Storia Naturale Università di Firenze, sez. Zoologica, 'La Specola', Italy. Photo Joanna Ebenstein [16-17] (ALL) Wellcome Library, London [18L] Science Museum, London / Wellcome Images [18R] Wellcome Library, London [19L, 19CL] Wellcome Library, London [19CR, 19R] Science Museum, London / Wellcome Images [20] Deutsches Hygiene-Museum, Dresden. Photo David Brandt [22-23T] Museo di Storia Naturale Università di Firenze, sez. Zoologica, 'La Specola', Italy / Bridgeman Images [22-23B] Museo di Storia Naturale Università di Firenze, sez. Zoologica, 'La Specola', Italy. Photo Raffaello Bencini / Archivi Alinari, Firenze / Topfoto [24L] Museo di Storia Naturale Università di Firenze, sez. Zoologica, 'La Specola', Italy / Bridgeman Images [24R] Museo di Storia Naturale Università di Firenze, sez. Zoologica, 'La Specola', Italy. Photo Saulo Bambi - Museo di Storia Naturale / Florence [25L] Musée Condé, Chantilly [25R] Dea / A. Dagli Orti / Getty Images [26-27] (BOTH) Museo di Storia Naturale Università di Firenze, sez. Zoologica, 'La Specola', Italy. Photo Joanna Ebenstein [28L] Galleria degli Uffizi, Florence. Photo Scala, Florence - courtesy the Ministero Beni e Att. Culturali [28R, 29L] Galleria degli Uffizi, Florence [29R] Royal Collection Trust. Her Majesty Queen Elizabeth II [30-31] British Library, London [32L] Museo di Storia Naturale Università di Firenze, sez. Zoologica, 'La Specola', Italy. Photo Saulo Bambi - Museo di Storia Naturale / Florence [32-33C] Museo dell'Opificio delle Pietre Dure, Florence [33R] Wellcome Library, London [34-35] Museo di Storia Naturale Università di Firenze, sez. Zoologica, 'La Specola', Italy / Bridgeman Images [36L] The National Library, Rome. Marka / Alamy Stock Photo [36C, 36R] Wellcome Library, London [37] (ALL) Wellcome Library, London [38] Private Collection / Photo Christie's Images / Bridgeman Images [40-41] (ALL) Wellcome Library, London [42] (ALL) Royal Collection Trust. Her Majesty Queen Elizabeth II [43L] Galleria degli Uffizi, Florence. Gabinetto dei disegni e delle stampe. Archivi Alinari-archivio Mannelli, Firenze / Topfoto [43R] Metropolitan Museum of Art, New York. Purchase, Joseph Pulitzer Bequest, 1924 (24.197.2) [44-47] (ALL) Wellcome Library, London [48-49] (BOTH) Deutsches Hygiene-Museum, Dresden. Photo Volker Kreidler [50-51] (BOTH) Private collection. Courtesy Erasmus House, Brussels. Photos Paul Louis [52T] Wellcome Library, London [52C, 52 BOTTOM ROW] (ALL) Science Museum, London / Wellcome Images [53 TOP ROW] (ALL), 53C] Science Museum, London / Wellcome Images [53B] Wellcome Library, London [55TL, 55TR] Museo di Storia Naturale Università di Firenze, sez. Zoologica, 'La Specola', Italy. Photo Joanna Ebenstein [55CL] Josephinum, Collections and History of Medicine, MedUni Vienna. Photo Joanna Ebenstein [55CR, 55BL, 55BR] Museo di Storia Naturale Università di Firenze, sez. Zoologica, 'La Specola', Italy. Photos Joanna Ebenstein [56-57] Museo Delle Cere Anatomiche 'Luigi Cattaneo', Bologna, Italy. Photo by Joanna Ebenstein [58] Muséum national d'Histoire naturelle (MNHN), bibliothèque centrale, Paris, Dist. RMN-Grand Palais / image du MNHN, bibliothèque centrale [59T] Photo The John Deakin Archive / Getty Images [59B] Valentin-Karlstadt-Musäum, Munich [60-61] (ALL) Josephinum, Collections and History of Medicine, MedUni Vienna. Photo Joanna Ebenstein [62-63] (ALL) Science Museum, London / Wellcome Images [64-65] Collezione delle Cere Anatomiche di Clemente Susini, Cagliari, Cittadella dei Musei, Italy. Courtesy Alessandro Riva. Photo © Università degli Studi di Cagliari [66] Santa Maria della Vittoria, Rome. Photo Joanna Ebenstein [68-69] Carmen Alto, Oaxaca, Mexico. Photo Joanna Ebenstein [70L] The Museum of Witchcraft & Magic, Boscastle, Cornwall [70C] Campion Hall Collections, Jesuit Institute, Old Windsor, Berkshire [70R] The Museum of Witchcraft & Magic, Boscastle, Cornwall [71L] Private collection [71R] Kupferstichkabinett, Berlin [72L] British Museum, London [72R] Vorderasiatisches Museum, Staatliche Museen zu Berlin [73 FIRST ROW L] Musée du Louvre, Paris [73 FIRST ROW CL, CR, R] Egyptian Museum, Cairo [73 SECOND ROW L] Egyptian Museum, Cairo [73 SECOND ROW CL] Detroit Institute of Arts [73 SECOND ROW CR] Musée du Louvre, Paris [73 SECOND ROW R] J. Paul Getty Museum, Los Angeles [73 THIRD ROW L, CL] Württembergisches Landesmuseum, Stuttgart [73 THIRD ROW CR] British Museum, London [73 THIRD ROW R] Kunsthistorisches Museum, Vienna [73 FOURTH ROW L] Museo Archeologico, Florence [73 FOURTH ROW CL] Egyptian Museum, Cairo [73 FOURTH ROW CR, R] British Museum, London [73 FIFTH ROW L] British Museum, London [73 FIFTH ROW CL] J. Paul Getty Museum, Los Angeles [73 FIFTH ROW CR] Cleveland Museum of Art [73 FIFTH ROW R] Egyptian Museum, Cairo [73 SIXTH ROW L] Manchester Museum [73 SIXTH ROW CL] Egyptian Museum, Cairo [73 SIXTH ROW CR] Royal Scottish Museums, Edinburgh [73 SIXTH ROW R] Petrie Museum, University College, London [74] Musée gallo-romain de Fourvière, Lyon [76] (BOTH) Metropolitan

Museum of Art, New York. Bequest of Susan Vanderpoel Clark, 1967 (67.155.23) [77] (ALL) Photo akg-images / De Agostini Picture Lib. / Veneranda Biblioteca Ambrosiana [79] Chiesa del Gesù Nuovo, Naples. Photo Joanna Ebenstein [80L] Private collection [80R] Photo © Carlos Olimpio Rocha [81] (ALL) Private collection [84] *Le Petit Moniteur illustré*, 1st November 1894 [85] (ALL) *Archivio*, Volume 33, R. Società Romana di Storia Patria; Deputazione romana di storia patria. University of Toronto, Robarts Collection (AAK-8109) [86-87] Cimitero delle Fontanelle, Naples. Photo Joanna Ebenstein [88] (BOTH) Wellcome Library, London [89L] Wellcome Library, London [89R] Photo © Archivio dell'Arte - Luciano Pedicini [90-91] Rijksmuseum, Amsterdam [92L] Collection of Tracy Hurley Martin [92R] Wellcome Library, London [93] Galerie d'Anatomie, École nationale supérieure des Beaux-Arts de Paris. Photo Joanna Ebenstein [94-95] UCLA Library, Los Angeles. QL61.R985ta 1710 [96-103] (ALL) Museo di Storia Naturale Università di Firenze, sez. Zoologica, 'La Specola', Italy. Photo Saulo Bambi - Museo di Storia Naturale / Florence [106-109] (ALL) Museo di Palazzo Poggi, Universita' di Bologna. Photo Fulvio Simoni [110-111] Museo di Palazzo Poggi, Universita' di Bologna. Photo Joanna Ebenstein [112-115] (ALL) Museo di Palazzo Poggi, Universita' di Bologna. Photo Fulvio Simoni [116-117] (ALL) Museo di Palazzo Poggi, Universita' di Bologna. Photo Joanna Ebenstein [118] Münchner Stadtmuseum, Sammlung Puppentheater / Schaustellerei, Munich [120-121] Collection Family Coolen, Antwerp [122L] Archivio Di Stato, Bologna. Photo akg-images / De Agostini Picture Lib. [122R] Wellcome Library, London [123L] Teylers Museum, Haarlem (TvB G 5681) [123R] Münchner Stadtmuseum, Sammlung Puppentheater / Schaustellerei, Munich [124] Museum of the History of Medicine of Catalonia, Barcelona [126TL] Münchner Stadtmuseum, Sammlung Puppentheater / Schaustellerei, Munich [126TR] Collection of Per Simon Edström [126BL] Collection of Enric H. March [126BC, 126BR] The Library Company of Philadelphia [127TL] Collection of Enric H. March [127TR] Prints & Photographs Division, Library of Congress, Washington, D.C. (LC-DIG-pga-02302) [127BL] The Library Company of Philadelphia [127BR] Münchner Stadtmuseum, Sammlung Puppentheater / Schaustellerei, Munich [128-129] (ALL) Deutsches Hygiene-Museum, Dresden [130L] Valentin-Karlstadt-Musäum, Munich [130R] Detroit Institute of Arts Museum [131L] Collection of Per Simon Edström [131CL] Münchner Stadtmuseum, Sammlung Puppentheater / Schaustellerei, Munich [131CR] Collection of Per Simon Edström [131R] Münchner Stadtmuseum, Sammlung Puppentheater / Schaustellerei, Munich [132-135] (ALL) Courtesy Ryan Matthew Cohn. Photos Daniel Schvarcz [136 TOP ROW] (ALL) Münchner Stadtmuseum, Sammlung Puppentheater / Schaustellerei, Munich [136 CL, 136C] Collection of Enric H. March [136 CR, 136 BOTTOM ROW] (ALL) Münchner Stadtmuseum, Sammlung Puppentheater / Schaustellerei, Munich [138-139] (BOTH) Münchner Stadtmuseum, Sammlung Puppentheater / Schaustellerei, Munich [140L] Wellcome Library, London [140CL] Science Museum, London / Wellcome Library, London [140CR, 140R] Mary Evans Picture Library [141L] Deutsches Hygiene-Museum, Dresden. Photo David Brandt [141C, 141R] Université de Montpellier, collections anatomiques. Photos © Marc Dantan [142] (ALL) Museo de la Medicina Mexicana, Palace of the Inquisition, Mexico City. Photo Joanna Ebenstein [143-145] (ALL) Courtesy Ryan Matthew Cohn. Photos Daniel Schvarcz [146-147] (BOTH) Collection Family Coolen, Antwerp [148] Collection of Stefan Nagel [149L] Collection Family Coolen, Antwerp [149R] Cohen Media Group / Courtesy Everett Collection / REX Shutterstock [151-155] (ALL) Université de Montpellier, collections anatomiques. Photos © Marc Dantan [156] Courtesy the National Library of Medicine, Bethesda, Maryland, USA [157] Collection of Stefan Nagel [159] Musée Carnevalet, Paris / Roger-Viollet / TopFoto [160-163] (ALL) Université de Montpellier, collections anatomiques. Photos © Marc Dantan [165] Tate, London. Photo akg-images / Erich Lessing. © Foundation Paul Delvaux, Sint-Idesbald - SABAM Belgium / DACS 2016 [166-169] (ALL) Université de Montpellier, collections anatomiques. Photos © Marc Dantan [170L] Courtesy Laerdal [170C] Boston Public Library, Print Department [170R] Courtesy Laerdal [171L] Musée des Beaux-Arts de Rouen [171R] Tate, London [172] (BOTH) National Fairground Archive, University of Sheffield, UK [175] Madame Tussauds Archives, London. Photo Joanna Ebenstein [176-177] Münchner Stadtmuseum, Sammlung Puppentheater / Schaustellerei, Munich. Photo Joanna Ebenstein [178] Photo © Ferrante Ferranti [180L] Museo di Storia Naturale Università di Firenze, sez. Zoologica, 'La Specola', Italy. Photo Joanna Ebenstein [180C] Wellcome Collection, London. Photo Joanna Ebenstein [180R] Museo di Palazzo Poggi, Universita' di Bologna. Photo Joanna Ebenstein [181L] Museo di Storia Naturale Università di Firenze, sez. Zoologica, 'La Specola', Italy. Photo Joanna Ebenstein [181C] Deutsches Hygiene-Museum, Dresden. Photo David Brandt [181R] Museo di Storia Naturale Università di Firenze, sez. Zoologica, 'La Specola', Italy. Photo Joanna Ebenstein [183] Photo © Massimo Listri / Corbis [184] San Francesco a Ripa, Rome / Bridgeman Images [185] Mary Evans / Grenville Collins Postcard Collection [186T] Museo de Málaga [186B] Neue Pinakothek, Munich (14680) / Photo Peter Horree / Alamy Stock Photo [188-189] (ALL) Private collection [192] The Dahesh Museum of Art, New York (1995.104) [193] Metropolitan Museum of Art, New York. Gift of Louis C. Raegner, 1927 (27.200) [194] (ALL) Private collection [195L] Photo Fine Art Images / Heritage Images / Getty Images [195C] Photo Atelier Eberth / ullstein bild via Getty Images [195R] Private collection, courtesy Richard Nagy Ltd., London [196-199] (ALL) Private collection, courtesy Richard Nagy Ltd., London [200] Collection of the Muséum National d'Histoire Naturelle. Photo © Bernard Faye / MNHN [202] Collection of Evan Michelson. Photo Joanna Ebenstein [203] Deutsches Hygiene-Museum, Dresden. Photo Joanna Ebenstein [204-205] Illustration by Chris Taylor © Thames & Hudson Ltd., London [206] Philadelphia Museum of Art, Pennsylvania. Gift of the Cassandra Foundation, 1969. © Succession Marcel Duchamp / ADAGP, Paris and DACS, London 2016 [208-209] (ALL) On Loan to the Hamburg Kunsthalle, Hamburg, Germany / Bridgeman Images. © ADAGP, Paris and DACS, London 2016 [211] Courtesy White Cube, London. Photo Stephen White. © Jake and Dinos Chapman. All Rights Reserved, DACS 2016 [212-213] (ALL) Photos © Stacy Leigh [214-215] Photo Koen Hauser. Hair and make-up Louise van Huisstede. Model Georgina Verbaan. Design BrandendZant. © Koen Hauser / UNIT CMA. [ENDPAPERS] Marbled paper, courtesy The British Library.

INDEX

Page numbers in *italic* refer to illustrations.

A Philosophical Enquiry into the Origin of Our Ideas of the Sublime and Beautiful (Burke) 203
Adam and Eve (Lelli) 107, *109*
agalmatophilia 188–9
Agnus Dei *70*, *71*
Albinus, Bernhardus Siegfried 37
Amazing Models (Hauser) 214–15
anatomical manikins 36, *52–3*
anatomical theatres 36, *90–1*, *92*, 122–3
Anatomical Venus
  Barcelona *124*
  Josephinum 33, *60–1*
  La Specola *14–15*, 18, *22–3*, *24–5*, *26–7*, 33, *34–5*, 180
  Medici Venus *14–15*, 18, *22–3*, *24–5*, *26–7*, 29, 32, 33, 36, 49, 61, 70, 107, 180, *181*, 185
  Pohl *10–11*, *176–7*, 203
  Spitzner 156, *161–3*
  Venerina *4–5*, *112–15*, 180
anatomists 186
Anatomy Act (1832) 130
artists' dissections 37, *42–3*
Average Americans (Leigh) 212–13

Bartoli, Daniello 96
bees *70*, *71*
Bellmer, Hans 208–9
Benedict XIV, Pope 18, 25, 97, 107
Bernhardt, Sarah 157, 158, *159*
Bernini, Gian Lorenzo *178*, 180–1, *183*, *184*
Bettini, Cesare *56–7*, 107
Biheron, Marie-Catherine 37
Birth of Venus (Botticelli) 28, *29*
Blancanieves (film) *149*, 156
Blessed Ludovica Albertoni (Bernini) *184*
blood circulation 37,
Bologna, Italy 18, 28, 106–7, 109–17
Bonnet, Louis-Marin 123
Book of the Dead 70
Botticelli 28, *29*
Bradbury, Ray 171
Brière de Boismont, Alexandre Jacques François 212
Burke & Hare 130
Burke, Edmund 203

cabinet of curiosities *32–3*
cadavers *see* corpses
Cagliari, Sardinia 36
Carnival dissections 123
Carter, Henry 48
Castan's Panopticum *131*, 132–4, *136*, *143–5*
Castle, Terry 203, 208, 210
Catholicism 18, 70–1, 76, 80, 106
Chapman brothers *211*
childbirth *116–17*, *144–6*, *151–5*, 156
Chovet, Abraham 37
Christianity 70–1, 77
Cardi, Ludovico (*Il Cigoli*) *42*, *43*
coffins *157*, *158*, *159*
Comstock Laws (1873) 171
corpses 28–9, *43*, 48–9, 81, 130, *186*, 203
corpus sanctus 66, 81
corsetry *129*, *132–3*
Counter-Reformation 80
Critique of Pure Reason (Kant) 209
Croix, François de la 97
Curtius, Philippe *8–9*, 175

da Vinci, Leonardo *42*, *43*
de Hoyos, Maria Elena Milagro 194
De Humani Corporis Fabrica (Vesalius) *1*, 37, *38*, *43*, 48
De Sade, Marquis 94, 96, 104–5, 188
death, attitudes to 122
death masks 59, 70, *74*, 170, *171*
Delvaux, Paul *164*, *165*
dermatological moulages *142*
Descartes, René 208
Desnoues, Guillaume 37, 96–7, 100–1
dissections 36, 37, *42*, *90–1*, 122–3, 130
Donnersmarck, Guido Henckel von 194
Duchamp, Marcel 206

écorchés *42*, *44*, *58*, 106–7, 110–11
Ecstasy of Saint Teresa (Bernini) 180–1, *183*
Egyptians 70, *72–3*
encaustic 70, *72–3*
Enlightenment 18, 24–5, 97, 203, 208–9, 213
eroticism 19, 122, 141, 170, 180–1
ethnographic busts 134
Eve, anatomical 36, *50*

ex-votos 70, *71*, *80*, *81*, 88, 212
eyes *56–7*, *64–5*

F is for Fake (Welles) 212
fashion mannequins 202
Ferrini, Guiseppe 28
fetal skeletons 89, 92, *93*, *94–5*
fetish objects 188–9
Florence, Italy 18, 25, 29, 81
Fontana, Felice 18, *24*, 28, 32
Francis of Assisi 71
French Revolution 33
Freud, Sigmund 202
Fuardi de Fossau, Joseph 53
'fugitive sheets' 36, *41*
Fuseli, Henry *130*, *141*, 157

Galen, Claudius 48
Gautier D'Agoty, Jacques-Fabien *44*, *46–7*, 48
genitalia *141*, *143*
Géricault, Théodore 170, *171*
Gérôme, Jean-Léon *192–3*
Given: 1. The Waterfall, 2. The Illuminating Gas (Duchamp) 206
Grand Tour 29, 32
grave robbers 130
Gray's Anatomy 48

Hammer, Emil Eduard 59, 130, *138–9*
Hauser, Koen 214–15
hermaphrodites *143*
Homo ex humo 92
Huijberts, Cornelius 89, 92, *94–5*
Hunter, William 37, 189, *190*, 194
Huysmans, J. K. 209

indulgences 80
ivory anatomies 18, *52–3*

Kant, Immanuel 209
Kokoschka, Oskar *195*, *197*
Krafft-Ebing, Richard von 188

La bella notomia (Cigoli) *42*
La Donna Scandalosa 88, *89*
La Poupée. Second Partie (Bellmer) 208–9
Lachmann, Esther (La Païva) 194
Laerdal, Asmund 170, *171*
Lecky, William 208
Leiden, South Holland 36, *90–1*, 92

Leigh, Stacy 212–13
Lelli, Ercole 106–107, *109–11*
Leopold II (1747–92) 18, 25, 28, 32
Life & Death Contrasted, or, an Essay on Woman 17
Life and Death 88
L'Inconnue de La Seine 170–1
London, UK 37, 130, 148
Luther, Martin 80
Lynch, David 171

Macabre Altar 89, *93*
Mahlar, Alma 195, *196*, 197, *198–9*
Manzolini, Anna Morandi 107
Mater gravida 36
materialism 208–9
medical museums 140, 194
Medici Venus *see* Anatomical Venus
memento mori 19, 36, 88, *89*, 92, 107, 108, 180
mezzotints 44, 48
Michelangelo 43
midwifery 52–3, *144–6*
see also childbirth
Millais, John Everett 171
miniatures 18
Montpellier, France 33
Moos, Hermoine 195, *196*, 197
morality 96, 123, 141
morgues 157, 170–1
Mori, Masahiro 204–5
moulages *142*, 149
mourning cult 171
Müller-Deym, Josef 37
mummification 70, *72*
murder pamphlets 130
museums *see also* panopticons
  advertising and catalogues 118, *122*, *123*, *126–7*, *136*, *138–9*
  anatomical 118, *126–7*, *131*, 140–1, 148, *149*, 171
  Palazzo Poggi 4–5, 18, 28, 97, 106–7, *109–15*
  La Specola, Florence 4–7, 25, 28–9, 32–3, *34–5*, 49, *55*, 60–3
  Hamburg *120–1*
  Josephinum 33, *60–1*
  Paris *58*, 149

Napoleon 33
natural philosophy 25, 28, 39
Neapolitan Cult of the Dead 77, *86–7*
necrophilia 189, 194
Newgate Calendar 130

Obscene Publications Act (1857) 171
obstetric phantom models *116–17*
Ophelia (Millais) 171

Padua, Italy 36, 43, 123
panopticons 59, *128–9*, 130, 131, *132–3*, *135*, 140, *143–5*

paper anatomies 16–17, 36, *41*
Paris Morgue 157, 170–1
Parker, Cornelia 171
passivity 19, 122, 213
Peron, Eva 194
Pinson, André Pierre 200
placebos 212
plague 77, 88, *93*, 97
plaster anatomies 12
Poe, Edgar Allan 174
Pohl, Rudolf 10–11, *128–9*, *141*, *176–7*
poppets 70
pregnancy 2, 20, 37, 147, 156
Protestantism 80
Psychopathia Sexualis (Kraft-Ebing) 188
purgatory 77, 80
Pygmalionism 189, 190–1, *193*

Raymond, Charles 188–9
'Real Doll' 195
religious statues 36, 68–9, 79, 80, 81, 88
reliquaries 76, 81
Resusci Anne 170, 171
Romans 70, 74, 75
Ruysch, Frederik 89, 92, *94–5*

Sacher-Masoch, Leopold von 188
'sacred representations' 81, *85*
saints
  beatification and canonization 106
  Counter-Reformation 80
  ecstasy 81, 180–5
  effigies 66, 68–9, 80–1, 88
  interventions 71, 88, 212
  relics 66, 76–7
Sandri, Giovanni Battista *116–17*
Santa Maria Nuova hospital, Florence 29, 33
Sarti, Antonio 140, 148
science as cultural construct 213
sex dolls 42, 212–13
sexual fetishes 188–9, *193*, 194–5,
sexual hygiene 141, *143*, 148–9
sexuality 184–5
shrines 71, 76, 80, 88
Siamese twins *141*
skulls 77
sleeping beauties 156–7, 164, *165*, 172
Sleeping Beauty (Curtius) 8–9, 156, *175*
Sleeping Venus (Delvaux) 164, *165*
Sleeping Venus (Spitzner) 156, *164*, *168–9*
spermatorrhea 141
Spitzner, Pierre 149, 150, *151–5*, 156, *160–3*, 164, *166–9*
statue fetishes 188–9, 190–1, *193*
Sue, Jean Joseph 92, *93*
supernatural ideas 184, 188, 202, 208, 213

Susini, Clemente 4–5, 14, 24, 29, 36, *64–5*
sword swallowing *141*
syphilis 96, *141*, *143*, 148, 149

tableaux 81, *85*, 88, 89, *93–7*
Tanzler, Carl 194
The Artist Sculpting Tanagra (Gérôme) *192*
The Flayed Angel (Gautier) *44*
The Nightmare (Hammer) *130*, 141
The Nightmare (Fuseli) *130*, 141, 157
The Rape of Prosperina (Bernini) *178*
The Uncanny Valley (Mori) 204–5
Theatre Rob Melich 148
'Theatres of Death' (Zumbo) 93–6, *96–103*, 104–105
Thérèse Raquin (Zola) 170
Titian 29
Tragic Anatomies (Chapman brothers) *211*
transcendence 184–5
travelling shows *126–7*
tuberculosis 166–7
Tussaud, Marie 33, *59*, 140
Tyburn gallows, London 130

Uffizi Gallery 29

van Butchell, Martin 189
Venerina see Anatomical Venus
Venus de' Medici 28, 29
Venus im Pelz (Sacher-Masoch) 188–9
Venus of Urbino (Titian) 29
Verrocchio, Andrea del 42–3
Vesalius, Andreas 1, 37, 38, 40, 43, 48
Vienna, Austria *60–1*
Von Hagens, Gunther 171
votives 70, 71, 80, 81, 82, 88, 107
voudou dolls 70

Wandelaar, Jan 40
wax
  anatomies 18, 28, 32–3, *60–3*, 96 *see also* Anatomical Venus
  characteristics 70
  figurines 70
  La Specola methods 48–9, *55*
  plaques 19
  religious associations 70–1, 78
  tableaux 81, *85*, 93, 96, 97, *98–9*, *102–3*
  types 49, 71
Welles, Orson 212
Who is Number One? (film) 185
wooden anatomies 48–9, *50–1*
Wunderkammern 25, 28, *30–1*

Zeiller, Gustav 2, 20, 181
Zoffany, Johann 29
Zola, Émile 170
Zummo, Gaetano Giulio (Zumbo) 93, 96–7, *98–9*, 100–1, *102–3*, 104–5

[ACKNOWLEDGMENTS]

THE ANATOMICAL VENUS would not exist without the many excellent scholars whose work has supported and inspired me throughout its creation, especially Philippe Ariès, Roberta Ballestriero, Maritha Rene Burmeister, Terry Castle, Eleanor Crook, Elizabeth Harper, Kathryn A. Hoffmann, Mel Gordon, Paul Koudounaris, Anna Maerker, Marta Poggesi, Vanessa Schwartz, and Marina Warner. Special thanks to Louise Baker, Claudia Corti, Rebecca Messbarger, and Michael Sappol for comments and fact-checking. Any errors are entirely my own.

This book was begun thanks to Amy Herzog, who invited me to do an essay on the topic for *WSQ* (formerly *Women's Studies Quarterly*). I am also indebted to Evan Michelson, who accompanied me on two trips to Italy to gather content; to Stefanie Rookis, who commissioned the exhibition that led to this project; and to the members of Neuwrite with whom I workshopped a chapter. Thanks are also due to Kate Forde and Ken Arnold, with whom I worked on the Wellcome Collection's 'Exquisite Bodies' exhibition in 2009.

This book was completed with the invaluable support of Tracy Hurley Martin, co-founder of the Morbid Anatomy Museum. I am also indebted to Laetitia Barbier, Eric Huang, and Joel Schlemowitz for assistance and advice; Eva Åhren, Heather Chaplin, Colin Dickey, Rachel Herschman, Wythe Marschall, Ronni Thomas and Friese Undine for research suggestions; and Richard Barnett, Eric Bleich, Ryan Matthew Cohn, Catherine Crawford, Caitlin Doughty, Marie Dauhiemer, Samuel Dunlap, Emily Evans, Megan Fitzpatrick, Tonya Hurley, Kate Koza, Chris Muller, Ilse Muñoz, Ceuci de Oliveira, Mark Pilkington, Cristina Preda, Amy Slonaker, Daisy Tainton, and Mike Zohn for moral and material support. I would also like to thank my family—Robert, Sandy, Donna and Laura Ebenstein and Judith and Dick Grose—for support of many kinds.

 At l'Université de Montpellier, special thanks to M Philippe Augé (*Président*); M Jacques Bringer (*Doyen de la faculté de Médecine*); Mme Caroline Girard (*Directrice de la culture scientifique et du patrimoine historique*); and Mme Françoise Olivier (*Chargée de la valorisation du patrimoine historique*).

Special thanks are also due to the knowledgeable and generous museum and library professionals Saulo Bambi, Sarah Bond, Katie Dabin, Florian Dering, Crestina Forcina, Phoebe Harkins, Selina Hurley, Cornelia S. King, Andreas Koll, Ross Macfarlane, Vanessa Toulmin, Manfred Wegner, Alfons Zarzoso. This book also relies on the generosity and vision of private collectors Paul Braithwaite, the Coolen family, Per Simon Edström, William Helfand, Enric H. March, and Stefan Nagel.

Many, many thanks are also due to my excellent editor Charlie Mounter and the brilliant team at Thames & Hudson—Tristan de Lancey, Jane Laing, Rose Blackett-Ord, Maria Ranauro, and Dan Streat—with whom this book is very much a collaboration.

[ABOUT THE AUTHOR]

JOANNA EBENSTEIN is a New York-based artist, curator and independent scholar. She is the founder of the Morbid Anatomy Blog and Library, as well as creative director and co-founder (with Tracy Hurley Martin) of the Morbid Anatomy Museum in Brooklyn. She is co-editor of *The Morbid Anatomy Anthology* (with Colin Dickey); co-author and featured photographer of *Walter Potter's Curious World of Taxidermy* (with Dr Pat Morris); and contributor to *Medical Museums: Past, Present, Future* (edited by Samuel J. M. M. Alberti and Elizabeth Hallam). Ebenstein acted as curatorial consultant on the Wellcome Collection's 'Exquisite Bodies' exhibition (2009) and has also worked with such institutions as Wellcome Collection, The New York Academy of Medicine, The Narrenturm Museum, and The Vrolik Museum. Her photography and writing have been exhibited and published internationally, and she lectures regularly around the world at a variety of popular and academic venues.

[MORBID ANATOMY MUSEUM]

MORBID ANATOMY MUSEUM was opened in Brooklyn, New York, in June 2014 by co-founders Joanna Ebenstein and Tracy Hurley Martin. The Museum is an expansion of Ebenstein's long-running Morbid Anatomy project, which began as a blog supporting her 'Anatomical Theatre' photographic exhibition. This exhibition, shown in 2007 at the University of Alabama at Birmingham's Museum of the Health Sciences, showcased the great medical museums of Europe and the United States and the curious artefacts they house, such as the Anatomical Venus. The Morbid Anatomy Museum hosts a lecture and event space, a café, and a gift shop, along with permanent and temporary exhibition space. Its permanent collection, The Morbid Anatomy Library, was founded in 2008 and makes available thousands of books, photographs, artworks, pieces of ephemera and artefacts relating to medical museums, anatomical art, collectors and collecting, the history of medicine, social attitudes to death, and curiosities. Most of these are drawn from Ebenstein's personal research collection. The Museum also hosts temporary exhibitions featuring objects drawn from private collections on topics such as mourning arts, waxworks in European panopticons, and more.

This book is dedicated to my Oma and Opa, Benno and Dina Ebenstein. With their great love of art, medicine and culture, I like to think they would have enjoyed this book had they lived to see it.

MORBIDANATOMYMUSEUM.ORG
MORBIDANATOMY.BLOGSPOT.COM